D1194483

An Introduction to Ion Exchange

Uniform with this volume and
in the same series:

An Introduction to Thermogravimetry (C. J. Keattch)
Basic Infrared Spectroscopy (J. H. v.d. Maas)
An Introduction to ^{19}F NMR Spectroscopy (E. F. Mooney)
Basic Laser Raman Spectroscopy (E. J. Loader)

An Introduction to Ion Exchange

by Russell Paterson
Chemistry Department,
University of Glasgow

PUBLISHED BY HEYDEN & SON LTD
IN CO-OPERATION WITH SADTLER RESEARCH LABORATORIES INC.

Published in Great Britain by
Heyden & Son Ltd.,
Spectrum House, Alderton Crescent,
London NW4

Co-published for exclusive distribution
in the USA by
Sadtler Research Laboratories Inc.,
3314–20 Spring Garden Street,
Philadelphia, Pa. 19104

Library of Congress Catalog Card No. 75–104789
SBN 85501 011 8

Made and printed in Great Britain at
the Pitman Press, Bath

Contents

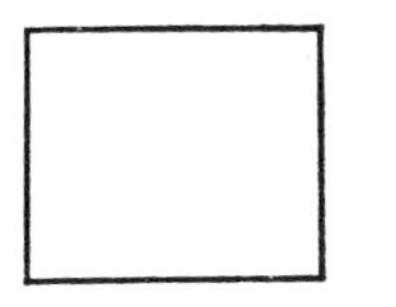

Foreword

The development of synthetic ion exchange resins by Adams and Holmes in 1935 laid the foundations for a new industry and a new unit operation which have grown remarkably since the war. The industrial value of ion exchangers, measured in terms of the processes in which they are used, is very great, and new uses are still being actively sought. New and improved ion exchangers continue to be developed; not only are methods of synthesising organic ion-exchange resins being continually improved, but there has been a resurgence of interest in inorganic ion exchangers, while the development of ion exchange membranes attracts considerable interest.

In parallel with the development of materials, and of processes based upon them, interest in the physical chemistry of ion exchange continues unabated. Such studies are of great fundamental interest, but they may also be relevant to the optimisation of industrial processes based on ion exchange. Studies of ion exchange membranes are of interest in many areas of biological and medical research.

The rapid growth in both theory and applications necessitates an up-to-date introductory text for those familiar with the basic principles of physical chemistry who wish to know something of this field. This book fulfils this purpose admirably; it will acquaint the reader with the outlines of the field and direct him towards further, more specialised reading.

AERE, Harwell
October 1969

C. B. AMPHLETT

1 Introduction

An ion exchanger is basically an electrolyte solution containing cations, anions and water, differing however, in that one or other ion is bound to an insoluble microporous matrix.

In the water-filled pores the remaining ion, of opposite charge to the fixed ion, is present in sufficient numbers to render the whole exchanger electrically neutral. These, the *counter ions*, are free to move through the matrix by diffusion or under electrical field and may be replaced by other counter ions (same sign) from solution, in the process of ion exchange. The material is a cation exchanger or an anion exchanger according as the counter ions are cations or anions respectively. The situation is summarised in Table 1.1.

TABLE 1.1

Exchanger Type	Fixed Charge	Counter Ion	Co-ion
Cation exchanger	anion (−)	cation (+)	anion (−)
Anion exchanger	cation (+)	anion (−)	cation (+)

The concentration of fixed charges in the exchanger, analogous to the concentration of an electrolyte solution, is known as the ion *exchange capacity*. It may be measured in a variety of units, the most common of which is milli-equivalents (of charge) per millilitre of wet resin. Since the swollen volume of wet resin is variable according to the counter ion type, it is usual to state this also. For example, the capacity of a typical general purpose cation exchanger is 2 mequiv/ml in the hydrogen form. Alternatively it is defined in terms of the scientific capacity; the number of equivalents of fixed charge per gram of dry matrix.

In the absence of absorbed salt molecules from the external solution the concentration of counter ion must equal the ion exchange capacity. Ions of the same sign as the fixed charge are known as *co-ions*, both in solution and when present within the exchanger pores.

Since electroneutrality is preserved at all times, the process of ion exchange is stoichiometric, e.g. one divalent counter ion will replace two univalent ions in the exchange process. A typical exchange in which calcium ion replaces hydrogen ion is shown below:

$$2\overline{H^+R^-} + CaCl_2 \rightleftharpoons \overline{Ca^{2+}R_2^-} + 2HCl \tag{1.1}$$

The barred species are in the exchanger phase and R^- represents one fixed charge on the exchanger matrix. Two points are to be noted. The first is that the reaction is reversible and will achieve equilibrium. A total conversion from the hydrogen form into the calcium form would require that the exchanger be equilibrated a number of times in the calcium chloride solution. The second is that the co-ion, in this case the chloride, is not involved in the process and so Eqn. (1.1) may be written more simply:

$$2\overline{H^+} + Ca^{2+} \rightleftharpoons \overline{Ca^{2+}} + 2H^+ \tag{1.2}$$

The co-ions will, however, affect the equilibrium if they form complexes or ion pairs with one or other of the counter ions in the solution, effectively reducing the concentration of that free ion and so its uptake into the exchanger. This effect is used in chromatography to improve the separation of ions in analysis and preparation techniques.

The position of the equilibrium is measured by the *selectivity coefficient*. This has the form of an equilibrium constant (to which it is related)

$$K_H{}^{Ca} = \frac{[\overline{Ca}] \cdot [H]^2}{[\overline{H}]^2 \cdot [Ca]}$$

where [ ] represent the concentrations of the respective ions. In the solution phase the usual unit of concentration is the gram ion/litre and in the exchanger, equivalent fraction. The equivalent fraction of an ion is the fraction of the total capacity of the exchanger attributable to that ion. If $K_H{}^{Ca} = 1$, the ion exchanger shows no selectivity; if greater than unity it will select calcium ions preferentially and if less than unity it will select hydrogen ions. The reasons for selectivity are discussed in Chapter 3. For general purpose organic resin exchangers the selectivity coefficient seldom exceeds three or four. In part, at least, the search for materials with higher selectivities has led to the development of the new inorganic ion exchangers.

The ion exchange process is in one sense not a chemical reaction but simply a rearrangement of existing species in solution and exchanger to achieve maximum chemical stability. No chemical bonds are broken or formed and there are no chemical products. It is therefore not surprising that the rate of the ion exchange process is determined by the rate of diffusion of the exchanging ions. The process occurs by diffusion through the aqueous pores of the exchanger structure. In the common organic resin exchangers these pores are considered to be of the order of 30 Å in diameter, although undoubtedly there are large holes or voids which may be some hundreds of ångströms across. The diffusion rates in the exchanger are usually some ten times smaller than in free solution, mainly due to the rather tortuous paths taken by the pores through the matrix. The term pore in this context is rather misleading. Most exchangers are made up of interlinked chains of organic polymer, lacking in regular internal structure and much like a knotted tangle of wool. (Fig. 1.1). The 'pores' are therefore the water-filled passages permeating the structure, random in shape and in dimensions. Since, in general, these structures are rather open, it is unusual for an ion to be excluded due to a size or ion-sieve effect. Notable exceptions

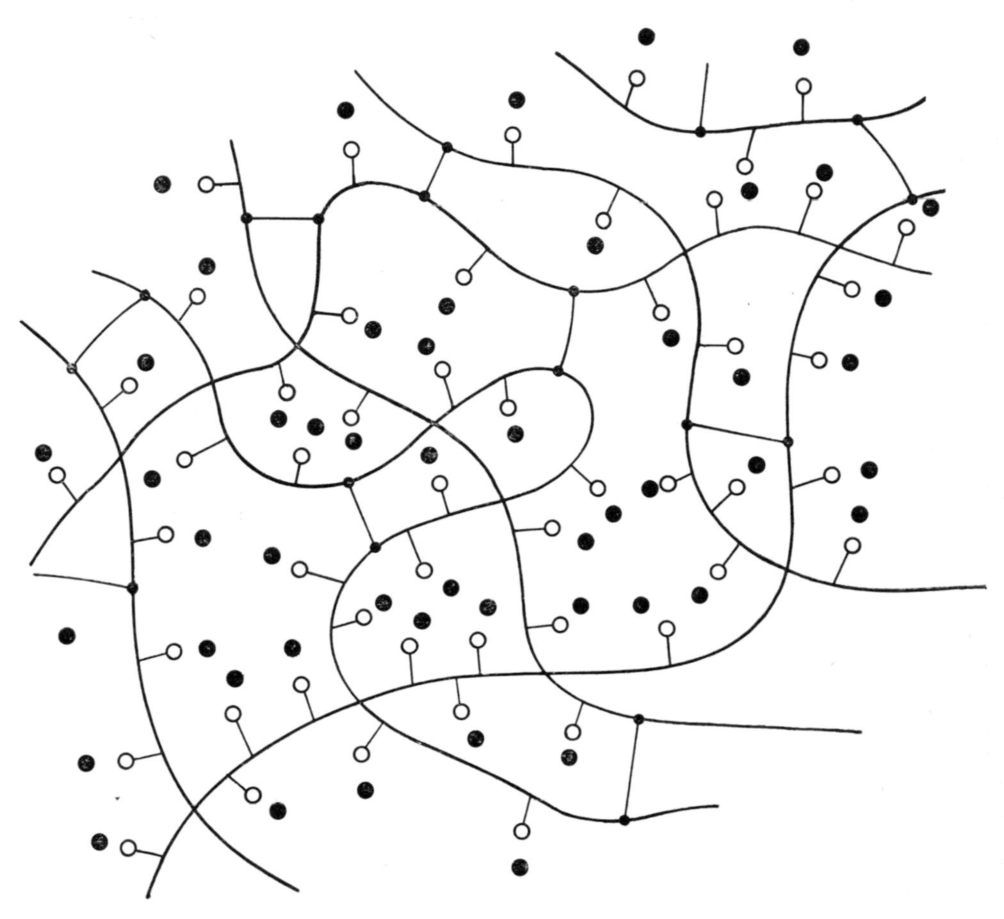

FIG. 1.1. Schematic representation of a modern ion exchange resin. The structure is based upon a random three-dimensional structure of linear polymer chains which are cross-linked at intervals along their length

——— linear polymer chains
●———● cross-linking bonds between chains
● counter ions
——○ fixed ionic groups

are found amongst the aluminosilicate exchangers, which have well-defined crystalline structures.

As membranes, ion exchangers have a number of remarkable properties, mainly due to their ability to exclude co-ions. One consequence of this is that, in a given ionic form, the counter ion is effectively the only mobile ion. As a result the transport number of the counter ion is unity. Electrodialysis employs this basic principle. In the cell, Fig. 1.2, a number of cation and anion exchangers are placed in series, each separated from the others by salt solution. Since the cation exchangers are permeable to sodium ion only, and the anion exchangers only to chloride, electric current causes depletion of salt in alternate chambers and enrichment in the others. In this way brackish water may be de-ionised or a valuable salt solution concentrated.

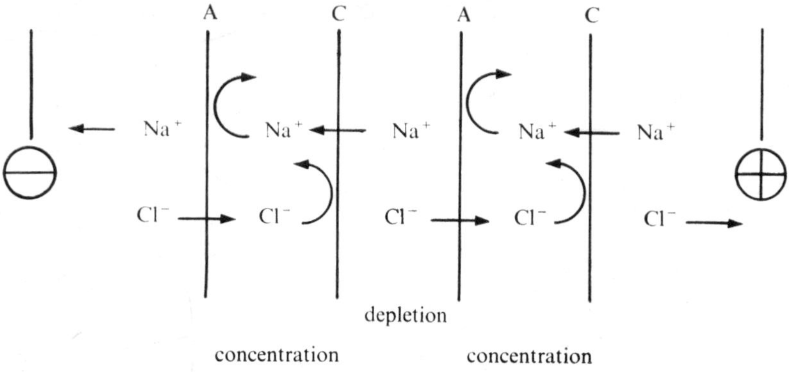

FIG. 1.2. Schematic representation of an electrodialysis cell made up of alternate cation, C, and anion, A, exchange membranes arranged in series

2 Ion exchange materials

Ion exchange was discovered in 1850 by Way and Thomas, two agricultural chemists who discovered the base exchange of soils. They found that a sample of soil, on being treated with an ammonium salt solution, took up the ammonium ion and lost an equivalent amount of calcium ion. It is not surprising that, at a time when the ionic nature of electrolyte solutions was not recognised, no coherent theory of the phenomenon was immediately available.

This cationic (or base) exchange, as it was called, was later traced mainly to the inorganic fraction of the soil and in particular to the clays and zeolites. A further contribution is made by the humic acid component, which is the residue of decayed organic matter. This is rich in hydroxyl and carboxyl exchange groups. The ion exchange capacity of the soil is in itself a very important factor in soil fertility, allowing it to store mineral ions, essential for plant growth, in an easily assimilable form. This aspect of soil chemistry alone is the basis for many hundreds of papers per year.

2-1. ALUMINOSILICATE EXCHANGERS

Silicate minerals are based upon SiO_4 units, in which the four oxygen atoms are tetrahedrally arranged around a central silicon atom. There are a number of different forms in which these silicate units may be arranged, each arrangement corresponding to a specific type of mineral (see Fig. 2.1). A few silicates contain simple discrete orthosilicate anions (SiO_4^{4-}) but the vast majority are polymeric.

If each unit shares two of its oxygens with its nearest neighbours, single-strand chains of composition $(SiO_3^{2-})_n$ are formed. Such minerals are known as pyroxenes and are closely related to the amphiboles, which also contain infinite chain anions, only this time arranged in a double chain of composition $(Si_4O_{11}^{6-})_n$. The silicate backbone of these minerals is negatively charged and this excess of charge is balanced to neutrality by the presence of cations. These cations may be of almost any type, although in natural deposits the minerals are named according to their cation composition. Examples of pyroxenes are enstatite, $MgSiO_3$; diopside, $CaMg(SiO_3)_2$; and spodumene $LiAl(SiO_3)_2$. This type of situation is quite common over the whole range of silicate minerals. In the minerals of this type the chains lie parallel, held together by the metal ions lying between them and are fibrous, cleaving easily in the direction of the long silicate chains. Asbestos minerals, which are amphiboles, typify this structure.

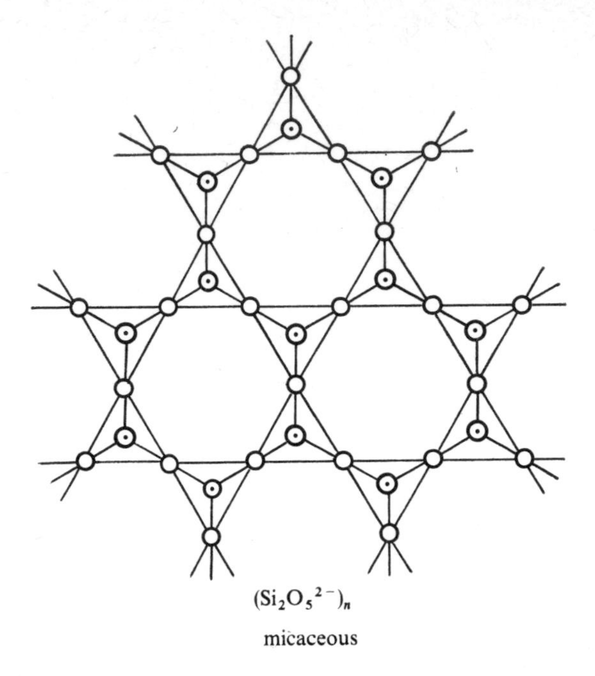

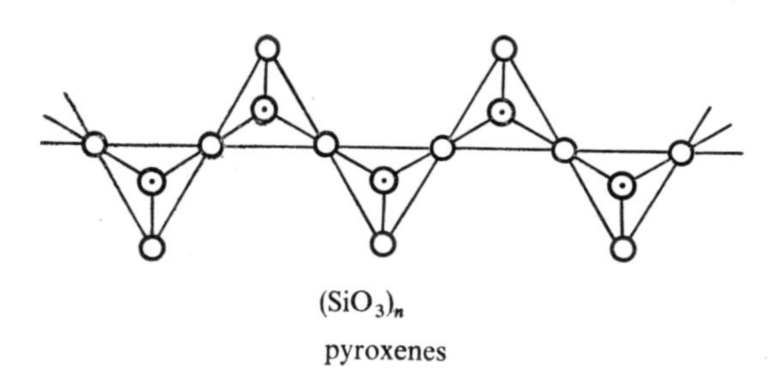

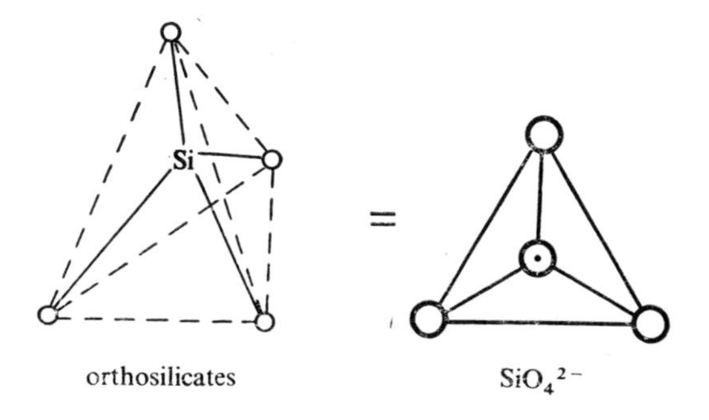

FIG. 2.1. The arrangement of silicate tetrahedra in common mineral types

When SiO_4 tetrahedra are linked by sharing three of their oxygen atoms with their nearest neighbours, infinite two-dimensional anions of general formula ($Si_2O_5^{2-}$) are formed. Many silicates have these sheet structures, with the sheets held together by the cations which lie between them. These minerals are typified by the micas and are readily cleaved along the direction of the silicate sheets.

Such minerals have little or no ion exchange application, for although they are insoluble and contain fixed negative charges on the polymeric silicate backbone, they lack the physical robustness which can only be achieved in a rigid three-dimensional structure. The next logical step in the series of silicate structures would be one in which each silicate tetrahedra shares all four of its oxygen atoms with its nearest neighbours. The formula of such a substance would be $(SiO_2)_n$; that is, the mineral silica itself. Silica is of course a neutral and uncharged solid and so of no possible use in ion exchange.* Pure silicate minerals are therefore of little use as ion exchange materials.

Aluminium may however replace silicon atoms in the silicates, isomorphously forming the family of aluminosilicates. Since aluminium has a nominal charge of 3+ and silicon 4+, each silicon so displaced leaves the silicate one charge more negative. It is therefore possible for aluminosilicates to be formed which combine a three-dimensional structure with, in addition, fixed anionic charges. Table 2.1 lists a number of common aluminosilicates with ion exchange interest.

These fibrous and lamellar minerals are prone to structural changes during reaction, in particular, the one-dimensional swelling and shrinking found in clays and micaceous

TABLE 2.1.[1] Aluminosilicates capable of Exchanging[a].

Fibrous	Lamellar	Three-dimensional
Zeolites	*Zeolites*	*Zeolites*
Edingtonite (390)	Stilbite (320)	Analcite[b] (450)
Natrolite (530)	Heulandite (330)	Harmotome (390)
Scolecite (500)		Levynite (490)
Thomsonite (620)	*Clays*	Mordenite[b] (230)
	Montmorillonite (100)	Chabazite[b] (400)
	Beidellite (100)	Faujasite[b] (390)
Clays		
Attapulgite (20–30)		
Sepiolite (20–30)	*Micaceous Alteration Products*	*Felspathoids*
	Glauconite (12–25)	Leucite (460)
	Illite (20–35)	Sodalite[b] (920)
	Vermiculite (100–150)	Nosean[b] (880)
		Ultramarine[b] (830)
		Cancrinite[b] (1090)

[a] The values in parenthesis are the approximate ion exchange capacities in mequiv per 100 g based on the hydrated formula weights.
[b] These have been investigated fairly fully as exchangers.

* Not so hydrous silica, see section 2.5.

alteration products. Least subject to these effects and so most useful in ion exchange are the zeolites and felspathoids of the third column of Table 2.1. These minerals have the required three-dimensional structure and have further more truly crystalline structures with all the regularity that this implies. As a result, their exchange behaviour may often be interpreted in terms of their internal matrix geometry which is, in many cases, available from X-ray diffraction data.

The typical building unit in many of the zeolites and felspathoids is the cubo-octahedral cavity, shown diagrammatically in Figs. 2.2 and 2.3. The open structure

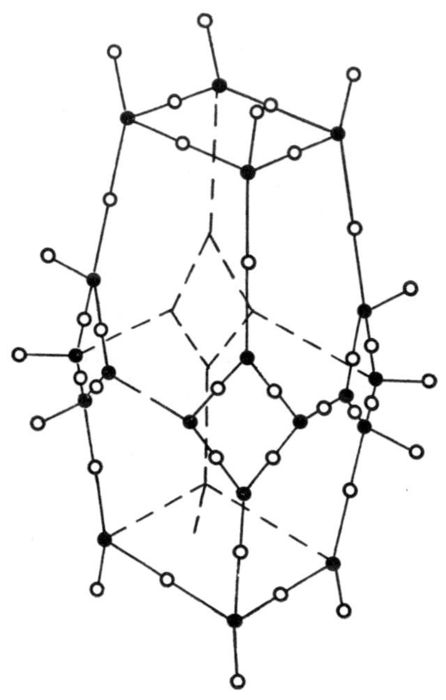

FIG. 2.2. The cubo-octahedral cavity typical of the basic structure of many zeolite exchangers. ● silicon, o oxygen

of these minerals makes possible their use as ion exchangers and molecular sieves. Each mineral has a characteristic and repetitive system of channels through which any diffusion process of ions or molecules must take place. The channel dimensions may vary periodically along their lengths. The minimum free diameters of these channels determines the lower limit for exclusion of ions or molecules. The maximum and minimum free diameters of the pores of the more important minerals are given in Table 2.2.

The complete regularity of the channels in these minerals and their 'molecular' size, ranging from 2 to 10 Å, has been exploited in their application as molecular sieves. Suitable minerals may be used to separate branched chain from normal chain aliphatic hydrocarbons simply on the basis that molecules with branched chains cannot enter the pores.

In aqueous solutions these channels will contain water molecules and anions, as well as cationic counter ions. If the relatively immobile carbonate or sulphate anions

TABLE 2.2. Maximum and Minimum Free Diameters of the Pores of Important Zeolites.

	Free Diameters (Å)	
	Maximum	Minimum
Faujasite	12·0	9·0
Linde Sieves	11·8	4·2
Chabazite	7·3	3·2
		2·2
Sodalite	6·6	2·2

are present they may well block channels to cation diffusion. There is also a range of ion sieve properties which depend on channel dimensions. As an example, neither ultramarine nor analcite will exchange with caesium ion, while chabazite and faujasite, having wider pores, will do so. Analcite has an even more striking property: it is able to completely discriminate between rubidium and caesium ions, by virtue of their crystal radii differing by only 0·2 Å.

Apart from the sources of naturally occurring deposits of these minerals, they have been synthesised by the slow crystallisation, under 'hydrothermal' conditions, from

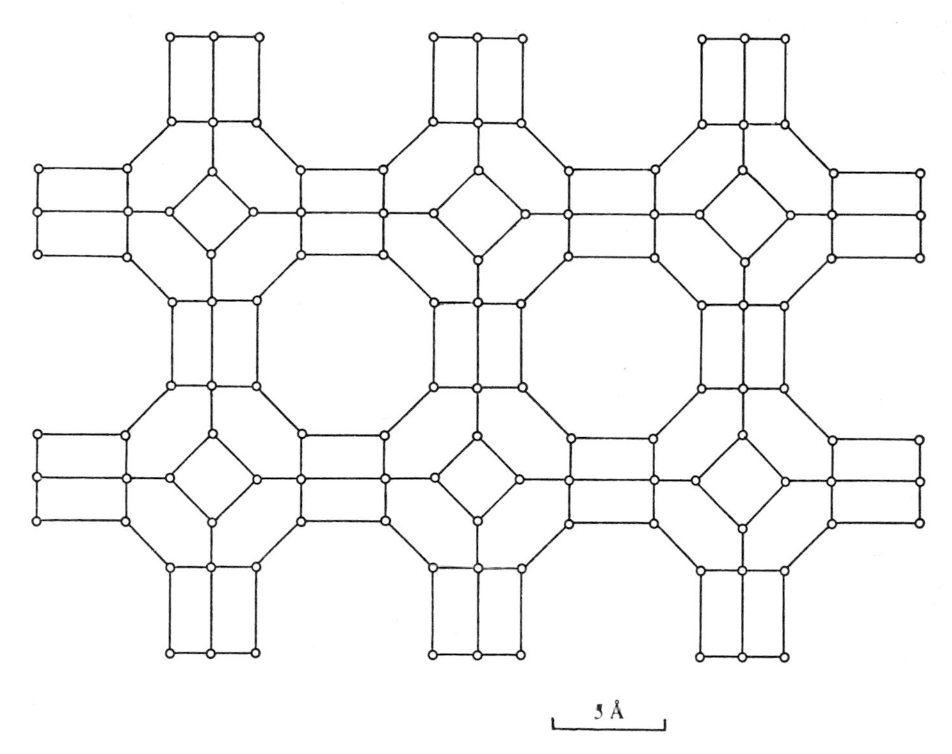

FIG. 2.3. The arrangement of cubo-octahedral cavities in the structure of Linde Sieve X – synthetic faujasite

sodium aluminosilicate gels of the correct composition. These synthetic minerals are identical to their natural counterparts and are commercially available as Linde sieves.

Aluminosilicate crystal exchangers are severely restricted for general use by their chemical instability in acidic and basic solutions and also by their low rates of ionic diffusion which may be more than four powers of ten less than in solution.

They only find a place in exchange techniques in specialised cases, where they may possess particular properties which are not available in the range of commercially available organic resin exchangers.

2-2. ORGANIC RESIN EXCHANGERS

The majority of these exchangers are based upon a matrix of cross-linked polystyrene. Upon this chemically and physically robust structure a very wide range of functional groups may be bound, each corresponding to a new type of exchanger. A degree of flexibility in the design of new exchangers is thus introduced, which in principle allows us to design particular exchangers for particular purposes.

Currently available ionogenic fixed groups range through sulphonate ($—SO_3^-$); carboxylate ($—COOH$); phosphinic ($—HPO_2^-$); phosphonic ($—PO_3^{2-}$); arsenic ($—AsO_3^-$); and selonic ($—SeO_3^-$) for cation exchangers to primary, secondary and tertiary amines; quaternary amines ($—NR_3^+$), phosphines ($—PR_3^+$) and tertiary sulphonium groups ($—SR_2^+$) for anion exchange. (In these representations the group R is an organic grouping such as the methyl group $—CH_3$).

Most of the current range of commercial exchangers are based on a matrix of polystyrene divinylbenzene copolymers.

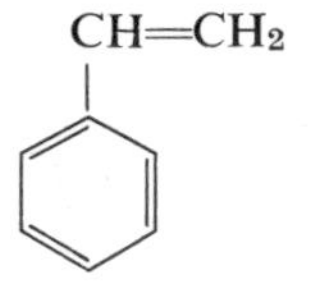

$CH{=}CH_2$ $CH{=}CH_2$

$CH{=}CH_2$

Styrene monomer 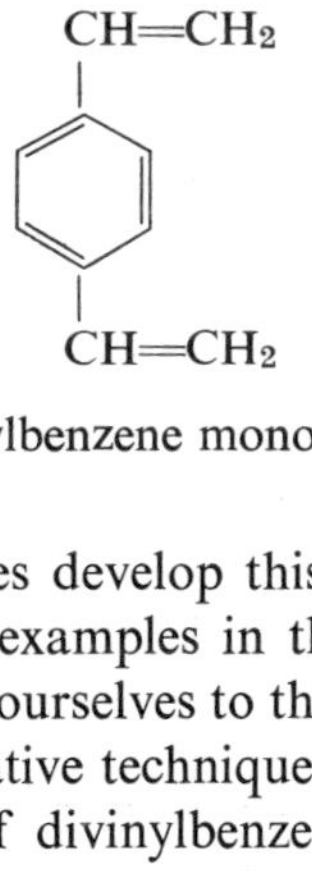 Divinylbenzene monomer

As polymer chemistry and preparative techniques develop this type of matrix may become less dominant – indeed there are many examples in the literature of other polymers as matrices. We shall, however, confine ourselves to this, the most common, in order to illustrate the principles of the preparative techniques.

Styrene is polymerised with a percentage of divinylbenzene which may range between 2 and 20%. The divinylbenzene (DVB) monomer, having two positions for polymerisation, cross-links the polystyrene chains, giving the required three-dimensional structure.

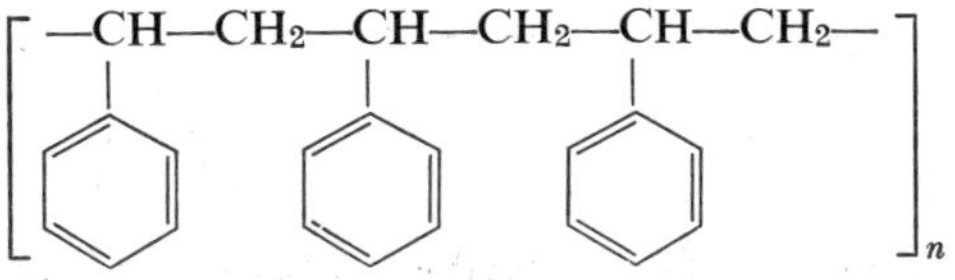

$[—CH—CH_2—CH—CH_2—CH—CH_2—]_n$

Polystyrene

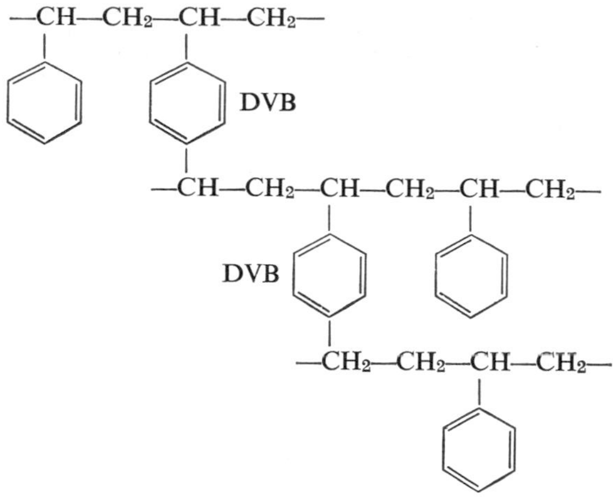

$—CH—CH_2—CH—CH_2—$

DVB

$—CH—CH_2—CH—CH_2—CH—CH_2—$

DVB

$—CH_2—CH_2—CH—CH_2—$

A copolymer of polystyrene and divinylbenzene, illustrating the effect of DVB in cross-linking the polystyrene chains to give a three-dimensional matrix

By a pearl copolymerisation process it is possible to obtain almost completely spherical beads in a wide range of particle sizes according to requirements. The general properties of the final exchanger are controlled by the functional group and by the degree of cross-linking.

The true degree of cross-linking is extremely difficult to determine by physical means in the final resin, but it is usually taken as the % DVB monomer in the original mixture used for polymerisation. The simple assumption is that each DVB molecule will provide one cross-link between polystyrene chains. We shall return to this problem of cross-linking in our discussion of selectivity in Chapter 3.

There is very good evidence that such polymers are usually not uniformly cross-linked throughout their volume. Even in the same single bead there may be regions of high and low cross-linking. Such resin exchangers may not be considered to be homogeneous nor to have pores in the same sense as the aluminosilicate crystal exchangers. They rather resemble a tangle of wool in which at irregular distances along the strands there are knots or cross-links, rendering the whole structure tight. This then is the basic matrix on which we must place functional groupings (Fig. 1.1).

Before considering the preparation and properties of such a matrix with a variety of functioning ionic groupings, we may summarise the situation. The modern ion exchange resin exchanger may not be considered as a homogeneous phase. Properties vary from bead to bead and from batch to batch of commercial exchanger of identical specification. However they are chemically robust and have well-defined physical characteristics such as swelling and selectivity with qualitative variations from batch to batch.

2-2a. Sulphonic acid exchangers

Dowex 50, Amberlite IR-120, Permutit Q, Duolite C-20 and others are of this type. The beads prepared by pearl polymerisation are sulphonated in sulphuric or chloro-sulphuric acid, giving practically complete monosulphonation of each benzene ring of the matrix. The resultant ion exchanger is not monofunctional as might be expected, since the sulphonate grouping may be placed on the phenyl ring of the styrene in one of three possible positions; the most probable being *meta* or *para*. The acidic strength of the sulphonate grouping of the exchanger will therefore vary considerably, being affected both by the position of the group in substitution and the specific environment of the group in the geometry of the matrix. The polystyrene sulphonic acid (often shortened to PSSA) exchanger resin is by far the most common cation exchanger, It will withstand repeated cycles of drying, swelling and de-swelling, will not decompose thermally below 120°C and is unaffected by exposure to strong acids, bases or redox solutions.

2-2b. Carboxylic weak acid exchangers

The second most widely used cation exchanger is the polymethacrylate exchanger. which contains as functional group the weak acid carboxyl grouping. It is prepared by copolymerisation of a mixture of acrylic or methacrylic acid with divinylbenzene or a similar compound with two vinyl groupings,

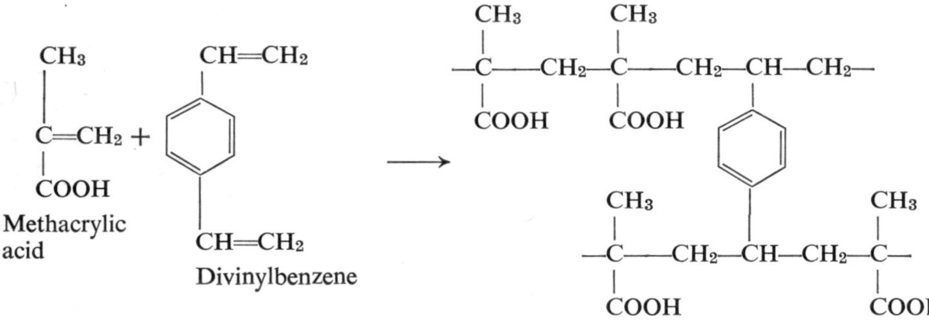

CH_3
$C{=}CH_2$ + $CH{=}CH_2$–C_6H_4–$CH{=}CH_2$ ⟶
$COOH$

Methacrylic acid

Divinylbenzene

$$-\underset{COOH}{\overset{CH_3}{C}}-CH_2-\underset{COOH}{\overset{CH_3}{C}}-CH_2-\underset{|}{CH}-CH_2-$$

$$-\underset{COOH}{\overset{CH_3}{C}}-CH_2-\overset{|}{CH}-CH_2-\underset{COOH}{\overset{CH_3}{C}}-$$

(the two CH groups linked through a benzene ring)

As with the case of the PSSA resin, the DVB is present in the polymerisation mixture to a minor degree (of the order of 10%).

The pearl polymerisation technique can be used to prepare spherical beads if, instead of the free acid, the carboxyl groups are first esterified. The esters may then be hydrolysed to give the free acid after polymerisation, Exchangers of this type commercially available are Amberlite IRC 50, Permutit H-70, Duolite Cs-101, Wolfatit CP-300.

2-2c. Anion exchange resins

Anion exchange resins are usually based on an amine or ammonium grouping as a source of positive fixed charge. Strong base exchangers are based upon quaternary ammonium groups ($-NR_3^+$), where R may be a methyl, ethyl or other organic substituents. Once again the polystyrene–divinylbenzene copolymer has proved the most usual matrix. The beads are chloromethylated and subsequently treated with ammonia or primary, secondary or tertiary alkyl amine.

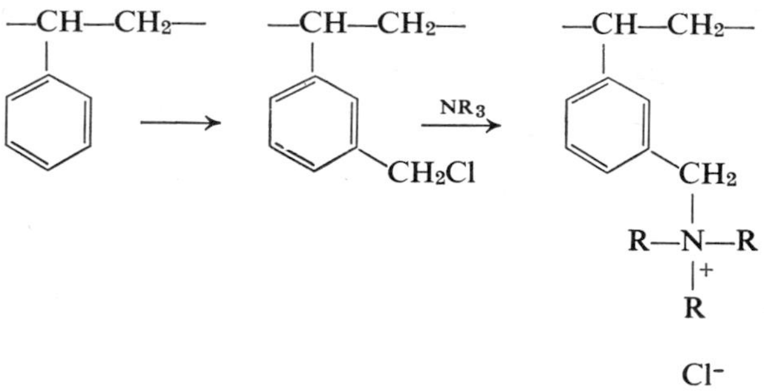

$$\text{—CH—CH}_2\text{—}(C_6H_5) \longrightarrow \text{—CH—CH}_2\text{—}(C_6H_4)\text{CH}_2\text{Cl} \xrightarrow{NR_3} \text{—CH—CH}_2\text{—}(C_6H_4)\text{CH}_2\text{—}\overset{+}{N}R_3 \; Cl^-$$

The chloromethylation step is a Friedel–Crafts condensation, catalysed by anhydrous zinc or stannous chloride, It has the complication that a side reaction may occur in which there is an elimination of hydrochloric acid from chloromethylated sites and adjacent hydrogen atoms, giving additional cross-linking.

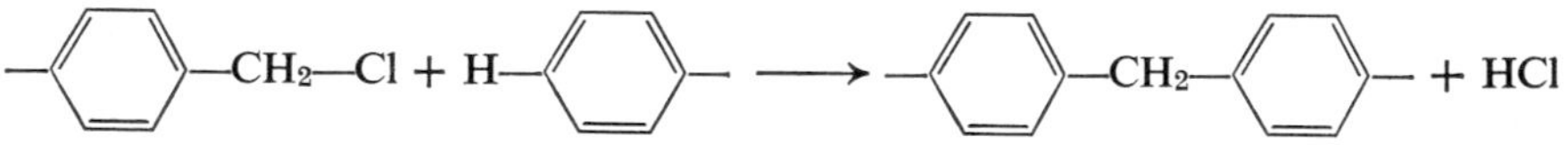

$$\text{—}C_6H_4\text{—CH}_2\text{—Cl} + \text{H—}C_6H_4\text{—} \longrightarrow \text{—}C_6H_4\text{—CH}_2\text{—}C_6H_4\text{—} + HCl$$

Phenyl adjacent

The common strong base ion exchangers are based either on trimethylamine or on diethanolamine, **1** and **2** respectively.

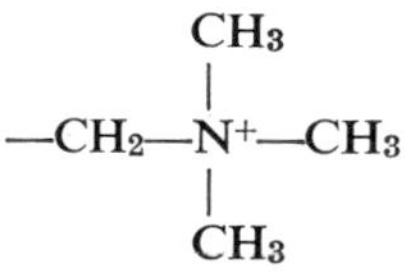

$$\text{—CH}_2\text{—N}^+(CH_3)_2\text{—CH}_3 \qquad\qquad \text{—CH}_2\text{—N}^+(CH_3)_2\text{—C}_2H_4OH$$

(1)	**(2)**
Dowex 1	Dowex 2
Amberlite IRA–400	Amberlite IRA–410
Permutit S–I	Nalcite SAR
Nalcite SBR	
Duolite A–42	
Permutit De Acidite FF	

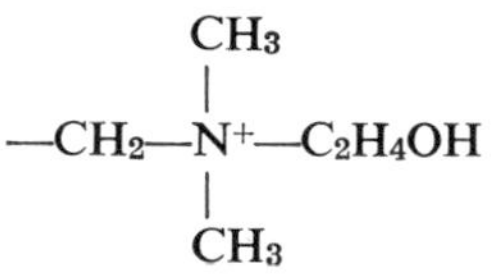

These strong-base exchangers are much less stable chemically and physically than their strong-acid counterparts. They are characterised by the fishy odour of amines, even at room temperature, and at 60°C will rapidly degrade, releasing tertiary amine (and methanol) by a Hofmann-type degradation.

It remains an outstanding problem to prepare a strong base exchanger with good chemical and thermal stability. Indeed, as will be seen in section 2.5, this was one of the reasons that the new inorganic exchangers were developed. The degree of success that has been achieved will be discussed there.

2-3. ION EXCHANGE MEMBRANE PREPARATION

A membrane is any plug, disc, rod or bead which may separate two solutions and still allow passage of some or all of the dissolved components.

Commercially available membranes are produced in sheet form. The aim is to produce thin membranes with good mechanical strength, flexibility and a high degree of permselectivity. The methods used to prepare exchanger particles are, in general, unsuitable for membranes. Sulphonation of thin cross-linked polystyrene sheets produces a brittle membrane with poor mechanical properties and a high degree of internal strain, which tends to crack. As a result of considerable effort and ingenuity, a number of membranes are now commercially available. They may be broadly classified as either homogeneous or heterogeneous.

For this purpose, a homogeneous membrane may be considered as one which is chemically uniform and so has an infinite three-dimensional structure. As with homogeneous beads, it may still contain irregularities of local cross-linking or of densities of fixed charges. Heterogeneous membranes are prepared from particles of ion exchanger of colloidal dimensions, which are then mixed with an inert adhesive binder and cast into sheets. Ion exchange is therefore a property of the exchanger particles and not of the supporting binder, which only provides the necessary flexibility and mechanical strength. The proportion of binder is conditioned by the opposing requirements of high exchange capacity and permeability and of good mechanical strength. In practice they usually contain around 50% of exchanger material. The available capacity and permeability depends upon achieving a maximum number of particle–particle contacts.

2-3a. Homogeneous membranes

These are the most difficult to prepare and yet are the most desirable, since their properties will be most reproducible and selectivities and exclusion should be greatest. The phenol–formaldehyde polymers provide such a matrix, and may be sulphonated or quaternised to form strong flexible homogeneous membranes. They have been extensively used in scientific studies, but have the disadvantage that the phenol group is in itself weakly acidic, making the exchanger bifunctional. By partial prepolymerisation of the styrene–divinylbenzene mixtures a more stable sulphonated membrane may be obtained. A membrane of this type is marketed by the Asahi Chemical Company of Japan. For industrial purposes mechanically strong membranes have been prepared by incorporating widely meshed plastic-fibre cloth. In general there remains (for commercial reasons) considerable lack of specific data on all such preparations.

2-3b. Graft co-polymers

More recently graft copolymers have been used to prepare membranes of outstanding mechanical strength and excellent electrical and exclusion properties.

Film mixtures of polyethylene and styrene or styrene–divinylbenzene are cross-linked in the solid state by exposure to γ-radiation from a cobalt-60 source. The ionising γ-rays cause disruption and random reformation of existing covalent bonds within the mixture, resulting in the formation of a strong three-dimensional matrix from the original linear components. From this matrix strong-acid or base exchangers may be prepared by standard methods. Commercially these membranes are available from the American Machine and Foundry Company as the AMF 60 and 100 series.

2-4. THE NEW INORGANIC ION EXCHANGERS

These are a varied class of insoluble inorganic precipitates, ranging from the non-stoichiometric and non-crystalline hydrous oxides of heavy metal ions to crystalline heteropolyacids.

They have arisen since the Second World War in numerous attempts, not all successful, to achieve two specific aims in ion exchange. The first, to have exchangers with spectacularly high selectivities for one ion or a group of ions that may then be used for their extraction and separation with great efficiency; the second, to develop exchangers capable of operation above the temperature limits of the available resin exchangers. These limits are relatively low: 150°C for sulphonic acid, and only 60°C for quaternary ammonium exchangers. In this field too, there is a requirement, specifically in the extraction and purification of radioactive fission products, that exchangers should be found which will survive high radiation doses which would destroy the exchange capability of organic resins. The preparation and function of the more common new inorganic exchangers will be followed in Chapter 7 by a discussion of their exchange chemistry, in which it will be possible to assess just how far these development aims have been fulfilled.

2-5. HYDROUS OXIDES

The hydrous oxides are the hydroxide precipitates of tri- and tetra- valent metal ions. Of those recorded the most common are:

$Al_2O_3 \cdot nH_2O$; $Fe_2O_3 \cdot nH_2O$; $MnO_2 \cdot nH_2O$; $SnO_2 \cdot nH_2O$; $ZrO_2 \cdot nH_2O$; $ThO_2 \cdot nH_2O$; and $SiO_2 \cdot nH_2O$.

To some extent these formulae are misleading. They are non-crystalline non-stoichiometric, glassy solids containing an indeterminate number of water molecules, n, which may be present as interstitial water or as hydroxide groups attached to the metal atoms. To prepare a hydrous oxide suitable for exchange, aqueous ammonia or other base is added to a solution of metal salt. The resulting gelatinous precipitate is washed filtered and dried at a temperature most usually between 25° and 100°C. There are of course wide limits and zirconia ($ZrO_2 \cdot nH_2O$) may be dried at up to 800°C and still retain exchange characteristics.[2]

Each oxide will have its own chemical peculiarities, but in general the gelatinous precipitate may be regarded as made up of random aggregates of metal atoms linked together by one or more oxygen atoms or by hydroxyl groups. Each metal atom may still have unshared hydroxyl or even water molecules coordinated to it. During the drying process a cross-linking occurs, due to the elimination of water molecules from adjacent hydroxyls, while at the same time the interstitial water is removed by evaporation. These two effects cause a hundred-fold shrinking of the precipitate volume and considerable mechanical strain within the glassy dried material. These strains are released when the material is placed in water. They break down quite vigorously with evolution of heat to give a finely granular material suitable for exchange purposes.

Since the original precipitates are gelatinous, and so of more or less random structure, there will remain in the dried material numerous metal atom sites which retain their metal hydroxide grouping intact. It is the presence of these sites which allows the hydrous oxides their ion exchange ability.

```
   |     |     |            |     |     |
 —M—O—M—O—M—          —M—O—M—O—M—
   |     |     |            |     |     |
  O H         OH            O           O
                    ——→     |           |
  OH          OH            |           |
   |     |     |            |     |     |
 —M—O—M—O—M—          —M—O—M—O—M—
   |     |     |            |     |     |
        HO                       HO
                               + 2H2O
```

The group M—OH may be intrinsically an acid or a base, indicating the possibilities of cation and anion exchange materials:

$$-\mathrm{M{-}O{-}H} = -\mathrm{M{-}O^-} + \mathrm{H^+} \qquad \text{acid}$$

$$-\mathrm{M{-}O{-}H} = -\mathrm{M^+} + \mathrm{OH^-} \qquad \text{base}$$

In fact all hydrous oxides are found to have cation-exchange characteristics in base and anion-exchange characteristics in acid bathing solutions.

Published capacity data for zirconia and stannic oxide are shown in Fig. 2.4.

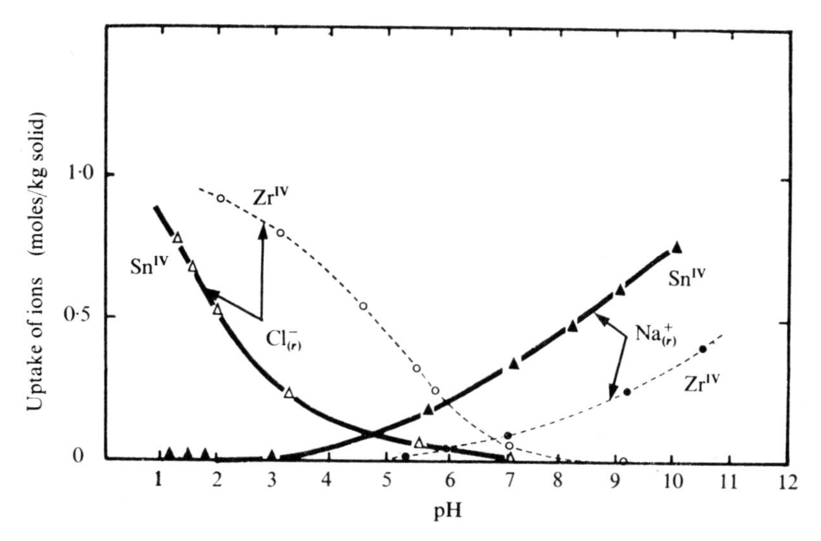

FIG. 2.4. The typical pH dependence of capacity on hydrous oxide exchangers. (Taken from reference 2, with permission)

The pH's of equal cation and anion capacities (the isoelectric points) reflects to a degree the inherent acidic and basic tendencies of the particular metal hydroxide. Stannic oxide is acidic and so retains cation exchange to regions of low pH; on the other hand, thoria is basic and an anion exchanger at pH's up to ten. It is most unlikely that these inversions of exchanger type are in all cases due to an amphoteric metal hydroxide group. One case where it is most probably true is in alumina:

$$\text{—Al—OH} \xrightarrow[\text{base}]{\text{NaOH}} \text{—AlO}^- + Na^+ \ (+ H_2O)$$

$$\text{—Al—OH} \xrightarrow[\text{acid}]{\text{HX}} \text{—Al}^+ + X^- \quad (+ H_2O)$$

For the other hydroxides with more well-defined acidic or alkaline tendencies there can only be postulated mechanisms in the absence of experimental evidence of mechanism for this pseudo-amphoteric behaviour:

(a) $\text{—M—OH} \xrightarrow{\text{NaOH}} \text{M—O}^- + Na^+ + H_2O$ normal acid

(b) $\text{—M—}(H_2O) \xrightarrow{\text{NaOH}} \text{—M}^-\text{—OH} + Na^+ + H_2O$ hydroxylated

(c) $\text{—M—OH} \xrightarrow{\text{HX}} \text{—M}^+ + (H_2O) + X^-$ normal base

(d) $\text{—M—OH} \xrightarrow{\text{HX}} \text{—M—}(H_2O)^+ + X^-$ protonated

It is possible that two other mechanisms exist, both involving sites which are occupied by coordinated water molecules: hydroxylation (b), and protonation (d). For zirconia with an isoelectric pH of 7 (indicating no inherent acidity or alkalinity in the hydroxide) the most probable mechanisms will be (b) and (d) respectively.

At the isoelectric point there is in the exchanger a situation where either there are equal numbers of positive and negative charges on the matrix or it is completely uncharged and the apparent capacities are due to sorption of external salt solution into the pores. The most probable explanation is the second, since these capacities are very small at the isoelectric point.

These remarkable materials can be used as simple ion exchangers in solutions of constant pH, although their selectivities for common ions are in no ways remarkable. Their chemical and physical stabilities vary widely over the range and may do so even from batch to batch of the same material prepared by identical procedures. The least soluble and most inert of the common hydrous oxides are zirconia and stannic oxide. Solubility tests sensitive to 10^{-7} molar in zirconium are negative for zirconia dried at 50°C and equilibrated for several weeks with water or dilute acid solution.

2-6. SALTS OF 12-HETEROPOLYACIDS

The ones commonly used in ion exchange are:

AMP	ammonium 12 molybdophosphate	$(NH_4)_3PMo_{12}O_{40}$
AMA	ammonium 12 molybdoarsenate	$(NH_4)_3AsMo_{12}O_{40}$
AWP	ammonium 12 tungstophosphate	$(NH_4)_3PW_{12}O_{40}$
QMP	quinoline 12 molybdophosphate	$(Q)_3PMo_{12}O_{40}$

They may be considered as salts of heteropolyacids of general formula

$$H_{8-n}^{+}(X^{n}Y_{12}O_{40})^{(8-n)-}$$

where Y may be molybdenum, tungsten, vanadium and others. X is the heteroatom, commonly phosphorus(V), arsenic(V), titanium(IV), zirconium(IV) or silicon(IV).

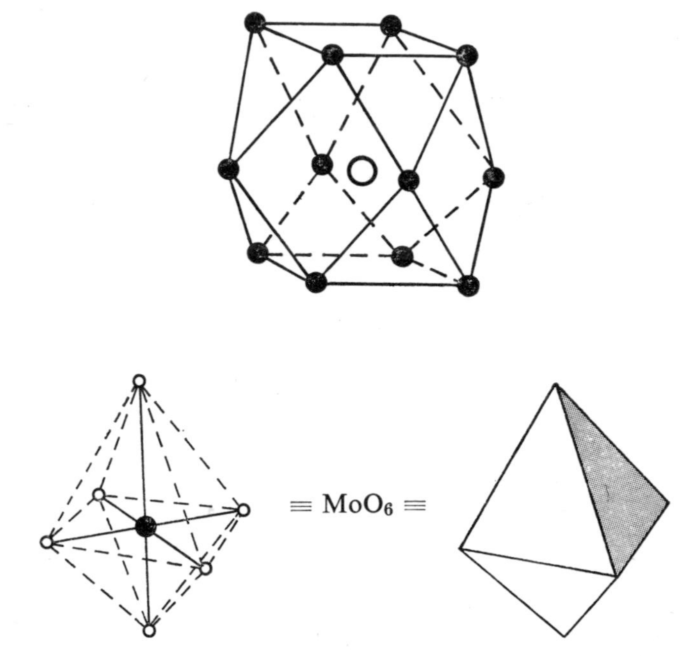

FIG. 2.5. (a) The chemical structure of the heteropoly-anion $(PMo_{12}O_{40})^{3-}$ present in AMP (oxygen atoms are omitted for clarity). (b) Octahedral arrangement of oxygen atoms around molybdenum

(The roman numerals indicate the oxidation number, n, of the elements.) The parent acid of AMP has therefore the formula $H_3(PMo_{12}O_{40})$. The heteropolyanion is based upon a structure of twelve MoO_6 octahedra clustered around, and sharing oxygen ions with, a central PO_4 tetrahedron. The molybdenum atoms are arranged at the vertices of a cubo-octahedron (Fig. 2.5) forming an almost spherical ion.

The acids themselves are very soluble in water but certain salts of the larger cations are insoluble, including NH_4^+, K^+, Rb^+, Cs^+, Tl^+, Ag^+, Hg_2^{2+} (mercurous ion), and with organic bases such as quinoline. These ions are sufficiently large to pack stably into the structure of heteropolyacid anions, lowering the lattice energy of the crystal sufficiently to render it insoluble in water. The crystal lattice also contains a large number of water molecules as water of crystallisation. Unlike the other exchangers discussed there is no three-dimensional matrix of covalently bonded atoms. Thistlethwaite[3] has shown, however, that of the three ammonium ions in AMP, only two are capable of exchanging with potassium ions from solution. The process of exchange is in some ways more akin to the formation of mixed crystals, since the exchanger is almost purely an ionic one.

Preparation. A typical preparation is that of Krtil and Kourim[4] who obtained AMP and AWP by the addition of excess of ammonium chloride to commercially available heteropolyacid in a medium of 0·1M hydrochloric acid. The precipitate was washed with 0·1M acid, absolute alcohol and dried at 40–60°C. In this case, analysis showed that some of the ammonium ions had been replaced by hydrogen in the preparation. The product is usually obtained as a very fine crystalline powder of mesh size ≈200 BSS, and capacity of approximately 1 mequiv/g.

2-7. ZIRCONIUM PHOSPHATE

Zirconium phosphate (ZrP) is one of the best known of the new inorganic ion exchangers. It has been prepared by many methods and the products obtained vary widely in crystallinity, capacity and selectivity. The truly crystalline compound with formula $Zr(HPO_4)_2 \cdot H_2O$ has been prepared by Clearfield and Stynes.[5] If solutions of soluble zirconyl salts and phosphoric acid are rapidly mixed at room temperature a white gelatinous precipitate is obtained. After washing by decanting and filtering a bulky gelatinous solid is obtained, which on drying shrinks and cracks to give a granular solid resembling the hydrous oxides, e.g. silica gel or hydrous zirconia. Further breakdown occurs when this solid is added to water until a stable product is obtained. The chemical composition of this amorphous gel is quite variable according to the specific conditions of precipitation. The phosphate to zirconium ratio is usually less than 1·7 (compared with 2·0 in the truly crystalline product).

ZrP is a fairly strongly acid exchanger, depending upon the acidic hydrogen ion of the bound phosphate for its exchange capacity. It has been successful in separations of fission products and has a particularly high selectivity for caesium ion, especially when it is a minor ionic component of the solution (section 7.2). The main restriction on its development for general purposes is its tendency to hydrolyse in solution, releasing phosphate. The phosphate release varies with the method used for the preparation, but is always serious in alkaline solutions of pH greater than 8 and is usually detectable at lower pH.

In the continued effort to obtain such new exchangers with possibly more spectacular selectivities and greater stability, almost every insoluble salt of tri- and tetra-valent metals has been investigated for exchange characteristics. A selection of metals and precipitating anions is given below. None are superior to ZrP or have less of a hydrolysis problem.

Sn, Zr, Ti, Th, Ce, Al, Cr, Ta and Bi.

Phosphate, arsenate, tungstate, chromate, silicate, antimonate, sulphide and oxalate.

2-8. FERROCYANIDE CATION EXCHANGERS

These are ferrocyanides of divalent ions which may be used as cation exchangers, and have a general formula $M_2^I(M^{II}Fe^{II}CN_6)\cdot nH_2O$, where M^{II} may be copper zinc or cobalt and others.[6]

Structural investigations indicate that there is a regular crystalline matrix which is cubic, based upon the octahedral arrangement of cyanide groups around the iron atoms, as shown in Fig. 2.6. This matrix is negatively charged. Overall electroneutrality is

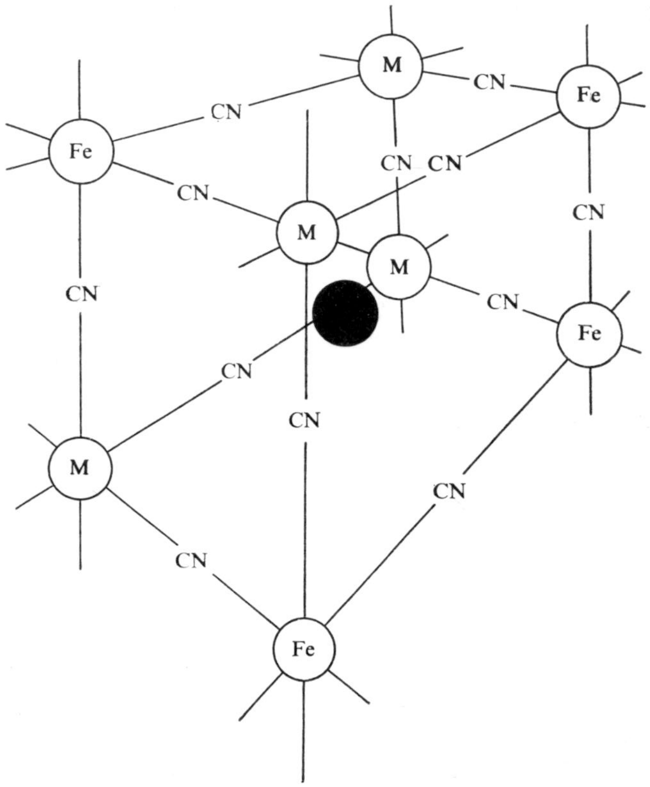

FIG. 2.6. The cubic structure of the unit cell of ferrocyanide exchangers, where M = Zn, Cu, Co and others. The exchangeable cation, ●, which is required for electroneutrality is situated within the central 'hole' in the crystal structure

achieved by the presence of cations (M^I) which are present within the open matrix structure. These cations are free to diffuse throughout the matrix and may be replaced by an ion exchange process.

Preparation. The exchanger is prepared by adding an excess of solution of precipitating metal salt to potassium or sodium ferrocyanide solution. The resulting precipitate is digested on a steam bath, filtered, washed and dried at 100°C.

REFERENCES

1. R. M. Barrer, *Proc. Chem. Soc.* **1958,** 99.
2. K. A. Kraus, H. O. Phillips, T. A. Carlson and J. S. Johnson, *Intern. Conf. Peaceful Uses of Atomic Energy, Geneva* **28,** 3 (1958).
3. W. P. Thistlethwaite, *Analyst* **72,** 531 (1947).
4. J. Krtil and V. Kourim, *J. Inorg. Nucl. Chem.* **12,** 367 (1960).
5. A. Clearfield and J. A. Stynes, *J. Inorg. Nucl. Chem.* **26,** 117 (1964).
6. V. Kourim, J. Rais and B. Million, *J. Inorg. Nucl. Chem.* **26,** 1111 (1964).

3 Equilibrium studies

The study of chemical equilibria is, of course, a study in thermodynamics. Rigorous thermodynamics, however, gives an inherently abstract treatment, devoid of the mechanical or microscopic images which would lead us to a feeling of greater intimacy and understanding of the phenomena. The most successful treatments have been based upon models which incorporate observed physical characteristics. In the case of organic exchangers, the most successful models take into account the swelling observed when these resins change environment or ionic form. Such treatments are careful to isolate as discrete terms pressure and volume contributions to the free energies of ions in the exchanger phase, defining them when otherwise they would simply contribute to the 'activity' coefficient; that sink for unknown parameters!

3-1. EQUILIBRIUM WITH SOLVENT (WATER)

Ion exchangers are able to absorb the solvent in which they are placed. This absorption is accompanied by the development of swelling pressure within the exchanger phase; particularly noticeable in organic resins where the matrix itself has a degree of elasticity and physically swells.

Organic resin exchangers are made up of linear polymer chains which have been cross-linked to give a three-dimensional network. Without cross-linking, such a polymer on immersion in water would simply dissolve to give a polyelectrolyte solution. The first water molecules entering the exchanger solvate the ionic groups within the matrix and as it proceeds the randomly arranged polymer chains unfold to allow for the greater bulk of the solvated ions. Counter ions and fixed ionic groups then constitute a very concentrated internal ionic solution. Consequently, as with any concentrated solution in contact with pure solvent, there is a tendency for ions to diffuse out of the exchanger and into the bathing solvent. Since one species of charge is fixed, only counter ions may freely diffuse. At the same time, external water molecules tend to be driven into the exchanger in an attempt to reduce the high concentration of the internal ionic solution of the resin phase. Dissolution of the exchanger is prevented by its cross-linking, but swelling remains as a result of the difference in concentration between resin and external solutions. This swelling pressure is the balancing force between the opposing tendencies of dilution of the internal solution and the rigidity of the exchanger matrix which tends to prevent such dilution. The situation is one analogous to osmotic equilibrium.

Eqn. (3.1) is derived in the Appendix.

$$\pi \tilde{V}_w = RT \ln \frac{a_w}{\bar{a}_w} \tag{3.1}$$

π is the swelling pressure, defined as the difference in hydrostatic between the internal solution of the exchanger and that outside, $\tilde{V}_w$ is the partial molar volume of the water molecule (18 ml/mole) and a_w and $\bar{a}_w$ the water activities in the external and exchanger phases respectively. The pressure volume term $\pi \tilde{V}_w$ compensates for the low activity of the water in the resin phase due to its high ionic concentration. Since the activity of water in electrolyte solutions is lowered by increasing concentration, when the exchanger is in equilibrium with more concentrated solution a_w is smaller and so π, the swelling pressure, is reduced. Thus an exchanger will be most swollen when in equilibrium with pure water solvent.

For a normal laboratory resin with a cross-linking of around 8% DVB swelling pressures of the order of 300 atmospheres may be developed within the exchanger when in equilibrium with pure water. The pressure volume term is therefore of importance in our understanding of ion exchange equilibria. On the basis of this simple model we may rationalise a number of phenomena:

(a) That polar solvents are, in general, better swelling agents than non-polar ones, due to their greater ability to solvate ions.
(b) That swelling is reduced in highly cross-linked polymers, due to increased rigidity (although the swelling pressure is higher).
(c) That swelling increases with capacity, particularly when the ionic components of the exchanger are known to show no serious tendency to ion pairing or complexing.
(d) That, for a given exchanger, swelling is reduced when the concentration of the external electrolyte is increased.

For a polystyrene sulphonic acid (PSSA) exchanger, the degree of swelling of the alkali metal forms is found to be in the order $Cs^+ < Rb^+ < K^+ < Na^+ < Li^+$, that is, in the order of their increasing solvated size. Thus, not the crystal radius, but the hydrated ion size is dominant in the swelling of this type of exchanger. If, however, there is severe restriction of swelling due to very extensive cross-linking or to a small crystal pore (zeolites), the full hydration of the internal counter ion may be impossible and this series may be partially or completely reversed. Weakly acid exchangers may also show a reversal of this swelling pressure series and Bregman,[1] investigating phosphonic acid exchangers containing the weakly acidic fixed group $—PO_3^{2-}$, found $K^+ > Na^+ > Li^+$. This and similar reversals in swelling order and their implied variation in swelling pressure are of fundamental importance in the understanding ion exchange selectivity. Indeed, as we shall see in section 3.5, this swelling reversal is paralleled by a selectivity reversal of these exchangers for these alkali metal ions. The exchanger tends to select the ion which will cause least swelling.

3-2. THE ORIGIN AND CONCEPT OF THE DONNAN MEMBRANE POTENTIAL

In the preceding section, we have considered the solvation and dissolution tendencies, which affect the exchanger as it comes to equilibrium with a solution. In this process ionic diffusion from the exchanger into the solution occurs only to a minute degree since it must be confined to the counter ion alone. Each counter ion which does leave the exchanger phase leaves behind an uncompensated charge on the matrix of the exchanger, making it more difficult to remove further counter ions, which now must do an increased amount of electrical work in escaping. Those which do diffuse out into the solvent remain near to the surface of the exchanger in a diffuse double layer. The separation of charge involved in this process sets up an electrical potential across the exchanger–solution interface: the Donnan membrane potential. This is illustrated diagrammatically in Fig. 3.1.

It must be stressed that very few ions indeed are sufficient to set up quite substantial voltages and also that the developed potential tends to return the counter ions of the

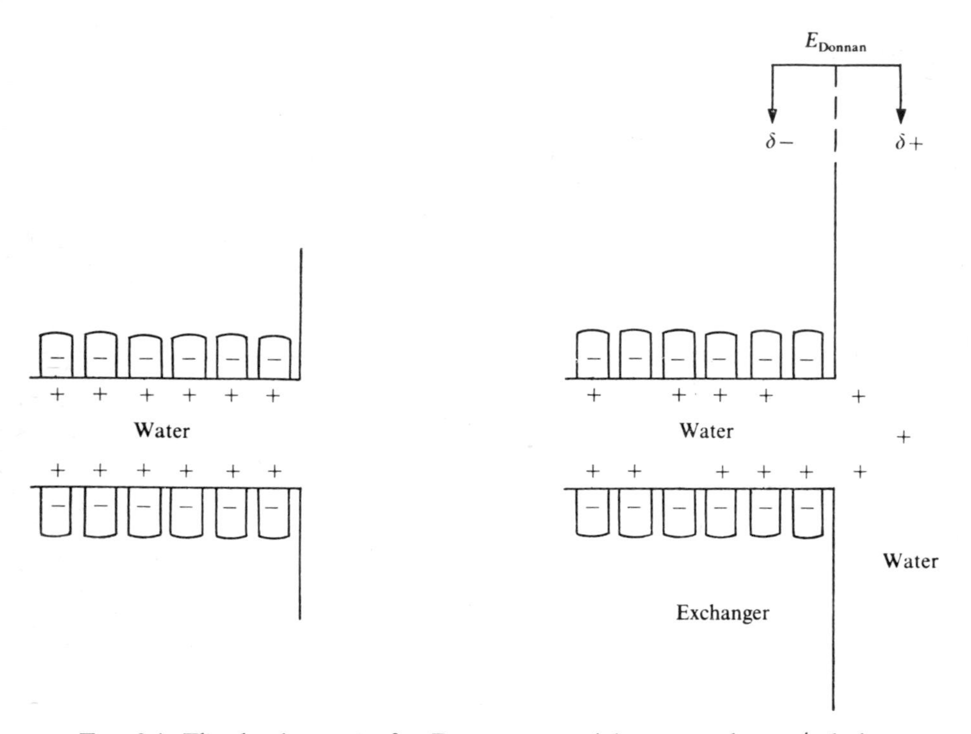

FIG. 3.1. The development of a Donnan potential at an exchanger/solution interface. *Left:* A hypothetical pore of a cation exchanger before immersion. *Right:* After immersion in water solvent

double layer to their matrix sites. The ions of the double layer are never a significant fraction of the exchanger capacity and so electroneutrality within the exchanger phase may be considered to be exactly maintained.

These very few ions and the resulting potential developed in their removal are of the utmost importance in our understanding of all ion exchange processes. Counter ions of the double layer are in dynamic equilibrium with those of the exchanger phase. Introducing foreign counter ions into the external solvent results in these mixing with and displacing the original counter ions of the diffuse layer, allowing them to enter and occupy sites within the exchanger.

3-3. ELECTROLYTE EXCLUSION AND ELECTROSELECTIVITY

The presence of the Donnan potential at the exchanger–solution interface also gives natural explanation to the fact that ion exchangers are permselective to their counter ions. In other words, an ion exchanger will virtually exclude all co-ions, allowing only counter ions into the exchanger phase.

Referring once again to Fig. 3.1, we see that should the co-ion (in this case the solution anion) approach the pore it would experience an electrical repulsion from the negatively charged matrix and so have little tendency to enter the pore volume. The Donnan potential automatically favours counter ion exchange and co-ion exclusion. Without further complication of our model we may consider the effect of valency upon ion selection and exclusion of an exchanger.

Since an electrical potential is defined as a force per unit charge, it follows that the larger the charge upon a co-ion, the more efficiently will it be excluded, while in the same manner counter ions of high valency will be preferred; the electroselectivity term. Co-ions entering the exchanger phase are accompanied by suitable counter ions to maintain electroneutrality and this amounts to electrolyte sorption.

From the arguments given, we may easily predict that for a strong acid (cation) exchanger, electrolyte exclusion will primarily be dominated by the valency of the co-ion, so that for the simple salts listed below the order of exclusion will be $Na_2SO_4 > NaCl > CaCl_2$. The effect of the valency of the cation counter ion is secondary but since the exchanger will select calcium over sodium the uptake of calcium chloride is greater than sodium chloride. In other words the valency of the counter ion will determine the order only when the co-ions of the salts involved are the same. Electrolyte exclusion is more efficient with salts where the co-ions are of high valency and the counter ions of low valency. We may further deduce that electrolyte or co-ion exclusion will

(a) Increase with increasing capacity
(b) Increase with decreasing solution concentration
(c) Decrease if co-ions form neutral or charged complexes with the counter ions of the solution.

3-4. SELECTIVITY OF AN ION EXCHANGER FOR COUNTER IONS; SELECTIVITY COEFFICIENTS AND EQUILIBRIUM CONSTANTS

When an ion exchanger is placed in a solution containing two counter ions A^+ and B^+ there will be an equilibrium set up between the exchanger and solution phases in which A^+ and B^+ are distributed to different degrees in each phase. We may represent this situation by a chemical equation.

$$\overline{R^-A^+} + B_{aq}{}^+ \rightleftharpoons \overline{R^-B^+} + A_{aq}{}^+ \qquad (3.2)$$

Unless specifically complexing, co-ions have no effect on this equilibrium other than indirectly upon the activity coefficients in solution. They may be neglected from the above equation. A thermodynamic equilibrium constant, K_{th}, may be written for this reaction.

$$K_{th} = \frac{a_{RB} \cdot a_A}{a_{RA} \cdot a_B} = \frac{[RB] \cdot [A]}{[RA] \cdot [B]} \cdot \frac{\gamma_{RB} \cdot \gamma_A}{\gamma_{RA} \cdot \gamma_B} \qquad (3.3)$$

In the general case, where the exchanging ions have valencies z_A and z_B, the thermodynamic equilibrium constant may be written

$$K_{th} = \frac{(a_{RB})^{z_A} \cdot (a_A)^{z_B}}{(a_{RA})^{z_B} \cdot (a_B)^{z_A}}$$

where a represents the activity of the species involved and is, in turn, equal to the concentration of that species multiplied by an activity coefficient such that, in general, for a species i

$$a_i = [i] \cdot \gamma_i \quad (3.4)$$

This expansion gives the thermodynamic equilibrium constant as the product of two terms, the first being a concentration and the second an activity coefficient quotient.

In experimental practice there is considerable use of the concentration term in Eqn. (3.3) which is termed the selectivity coefficient and given the symbol $K_A{}^B$ since it measures the tendency of the exchanger to select B over A. Thus:

$$K_A{}^B = \frac{[RB] \cdot [A]}{[RA] \cdot [B]} \quad (3.5)$$

If $K_A{}^B$ is greater than unity, the exchanger selects ion B. If it is less than unity, it selects ion A and if it is equal to unity then the exchanger shows no preference for either ion. Concentrations may be measured in molar, molal, or equivalent fraction units. To distinguish them the concentration units prefix the selectivity, e.g. molar, molal, and equivalent fraction selectivities.

Particularly in cases where chromatographic applications are involved, it is usual to represent the selectivity of the exchanger for the ion as an isotherm as shown in Fig. 3.2. Three simple and common examples of isotherms are given, these are linear I, convex II, and concave III. Defining our concentrations in terms of equivalent fractions, we see that the selectivity constant $K_A{}^B$ is now given by:

$$K_A{}^B = \frac{\bar{X}_B \cdot X_A}{\bar{X}_A \cdot X_B} = \frac{\text{Area bP}}{\text{Area aP}} \quad (3.6)$$

From this relationship we see that the selectivity constant is unity for a linear isotherm; there will therefore be equal ratios of the two ions in each phase. Equally it is obvious that curves of types II and III represent the cases in which B is selected over the whole range of solution and exchanger compositions, and the second in which ion A is similarly selected over B. If, in more complex cases, an isotherm is obtained which crosses the linear one, then the system shows selectivity reversal at the point of intersection. In the chapter on inorganic exchangers we will discuss one such case which arises in the exchange of caesium for hydrogen ions on zirconium phosphate.

3-5. THE DISTRIBUTION COEFFICIENT

Yet another common function used to express the position of an equilibrium is the distribution coefficient. Once again this is a function which is primarily of use in chromatographic theory. It is commonly defined as the ratio of the concentration of the ionic species in the exchanger phase to that in the solution. Concentrations are usually given in molar or molal units. The molal distribution coefficient, λ_i, of the ion i is defined as:

$$\lambda_i = \frac{\bar{m}_i}{m_i} = \frac{\bar{X}_i}{X_i} \cdot \frac{\bar{m}}{m} \tag{3.7}$$

m and X represent concentration in molality and equivalent fractions respectively, while barred terms refer to the exchanger phase. m_i and $\bar{m}_i$ are the molalities of the

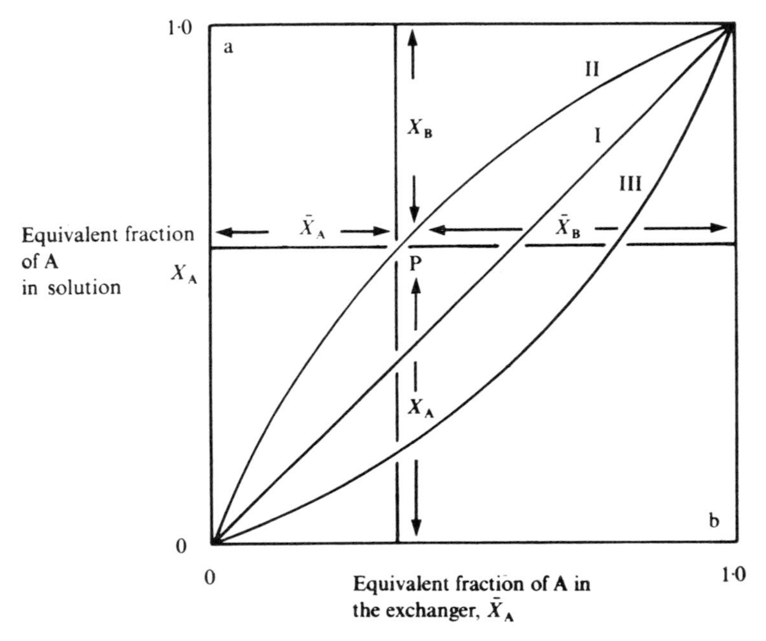

FIG. 3.2. Representation of ion exchange equilibrium as an exchange isotherm

ion i in the solution and exchanger phases, while m and $\bar{m}$ are the total molalities of the ions in each phase. If concentrations are measured in molarities a molar distribution coefficient is defined.

These coefficients are concentration sensitive, increasing with dilution of the solution. In general the distribution coefficient is of value as a practical guide to separation procedures in chromatography. For trace loading of A^+ on an exchanger predominantly in the B^+ form it assumes much greater interest. Under these conditions both the solution and exchanger are almost totally in the B-form, so that $\bar{X}_B \approx X_B \approx 1$. This corresponds to a region near the origin of the exchange isotherm.

$$\lambda_A = K_B{}^A \cdot \bar{m}_B/m_B \approx K_B{}^A \cdot \bar{m}/m \tag{3.8}$$

The approximation $m_B \approx m$ and $\bar{m}_B \approx \bar{m}$ is a good one. $K_A{}^B$ is effectively a constant since we are considering only a very short portion of the isotherm, so that a plot of log λ_A versus log m will be linear and of slope -1. For the general case where the ions have valency z_A and z_B the distribution coefficient takes the form

$$\lambda_A = (K_B{}^A)^{1/|z_B|} \cdot \left(\frac{\bar{m}}{m}\right)^{\frac{z_A}{z_B}} \tag{3.8a}$$

$|z_B|$ represents the positive (modulus) value of the valency. [In this case log λ_A versus log m will have slope $-(z_A/z_B)$.]

Kraus[2] has used this as a method to confirm that the uptake of ions by zirconia and zirconium tungstate is indeed due to ion exchange and not physical adsorption

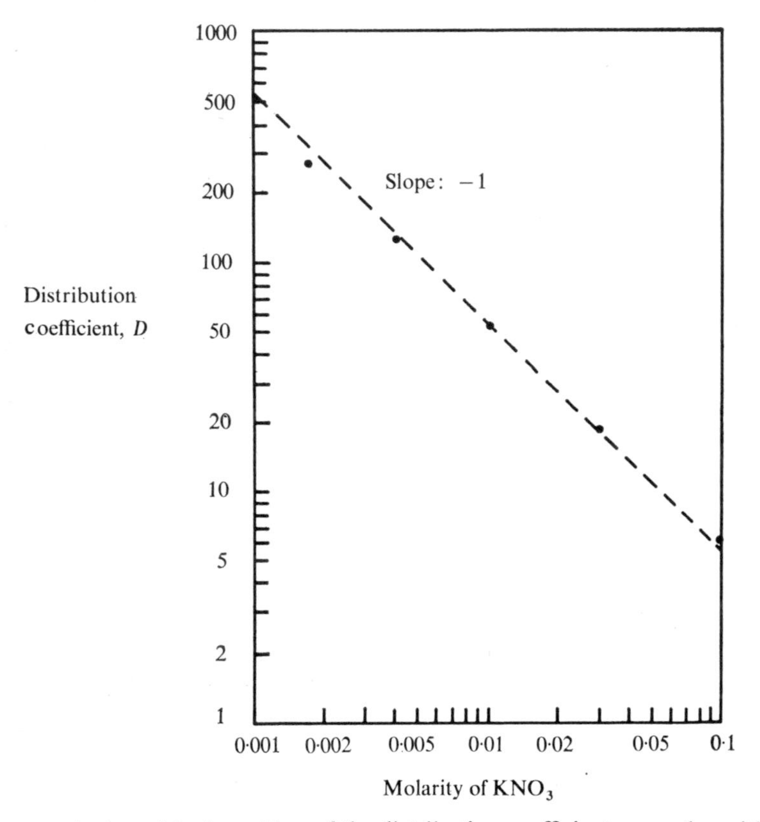

FIG. 3.3. A plot of the logarithm of the distribution coefficient versus logarithm of potassium nitrate concentration in an equilibrium between (tracer) bromide and nitrate on hydrous zirconia. The plot is linear, of slope -1, as predicted from Eqn. (3.8). (Taken from reference 3, with permission)

(Fig. 3.3). A more general use found for this approach is in the case of elution techniques, where two or more ions are to be separated by elution with a third which is the major component of the solution and the exchanger. (The rate at which ions move in chromatography is proportional to their distribution coefficient.)

3-5a. The effect of complex formation upon the distribution coefficient

If A is a cation in a solution containing the complexing ligand X, a series of complexes of the type AX, AX_2, AX_3, . . . AX_n will be formed in proportions dependent upon the concentration of X in solution and the equilibrium constants for the reactions (3.9).

$$
\begin{aligned}
A + X &= AX & \beta_1 &= m_{AX}/m_A \cdot m_X \\
A + 2X &= AX_2 & \beta_2 &= m_{AX_2}/m_A \cdot m_X^2 \\
&\cdots\cdots & &\cdots\cdots \\
A + nX &= AX_n & \beta_n &= m_{AX_n}/m_A \cdot m_X^n
\end{aligned}
\tag{3.9}
$$

β_n is the overall stability constant for the formation of the complex AX_n. It is large and positive when the complex is stable.

Under conditions in which A ion is present at trace level in a solution of BY, where Y is a non-complexing anion, Eqn. (3.8) holds and λ_A is a constant. m_A so defined is the concentration of free (uncomplexed) A ions. Experimentally λ_A is obtained by observing the radioactivity of the solution isotopically traced with A before and after equilibration with a standard resin sample. If ligand X, specific for A ions, is introduced into the solution the measured distribution coefficient will be that of Eqn. (3.10)

$$\lambda'_A = \bar{m}_A/m_A + m_{AX} + m_{AX_2} + \ldots m_{AX_n} \tag{3.10}$$

provided complex species are not absorbed by the resin. (This simple treatment is therefore confined to neutral or anionic complexes formed when X is a negatively charged ion.)

From Eqns. (3.8) and (3.9):

$$\lambda'_A = \lambda_A/1 + \beta_1 m_X + \beta_2 m_X^2 + \ldots \beta_n m_X^n \tag{3.11}$$

Since the denominator of this equation is large and positive, the result of complexing is to reduce the distribution coefficient for the ion complexed. This effect is used extensively to obtain improved separations in chromatographic analysis.

Re-arranging Eqn. (3.11) we obtain

$$\alpha = \left\{\frac{\lambda_A}{\lambda'_A} - 1\right\} / m_X = (\beta_1 + \beta_2 m_X + \beta_3 m_X^2 + \ldots)$$

If λ'_A is measured over a range of ligand concentrations from zero upwards, a plot of α versus m_X will give β_1 as m_X tends to zero. Taking this result for β_1, β_2 may be obtained from the function $(\alpha - \beta_1)/m_X$ as m_X tends to zero, and so on for the other equilibrium constants. This method of calculation of stability constants was first applied by the Swedish chemist Leden.

3-6. EXPLANATION OF ION EXCHANGE SELECTIVITY

In the following discussion of the factors which account for selectivity in ion exchangers we will confine our attention to series of ions belonging to the same group of the periodic table and thus having the same valency and electronic structure. In this way we may see more clearly the effects of ion size and hydration upon the equilibria, without the added problem of the influence of electronic and valency differences entering our discussions. In particular, we shall consider as models the alkali metal ions and the halogen ions which have the further property of inert gas configuration. Their interactions with other ions is confined to, at most, ion association in which no chemical bond is formed.

Other interactions do occur, of such specific types that their influence on selectivity is immediately obvious.

(a) Coordination of metal ions with fixed ionic groups undoubtedly occurs with carboxylic (weak acid) exchangers where there is the possibility of complex formation with, in particular, the transition metal ions. This effect is exploited in order to obtain even higher selectivities for such ions by introducing exchangers with specifically

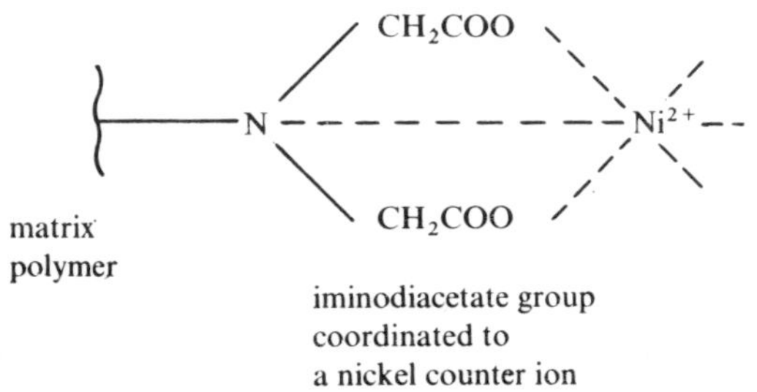

CH_2COO
N
Ni^{2+}
CH_2COO
matrix
polymer
iminodiacetate group
coordinated to
a nickel counter ion

FIG. 3.4. The complexing iminodiacetate anion present in Dowex A-1

complexing groups. Dowex A-1 is of this type containing iminodiacetic acid groups and is represented in Fig. 3.4 where it is shown how strong chelate complexes may be formed in each of these cases.

(b) London or van der Waals forces may also be decisive in determining the selectivity of an ion exchanger. These are non-ionic forces and often involve interactions with the matrix backbone itself. It has, for instance, been found that styrene-based resins prefer counter ions with aromatic groups to those with aliphatic groups. The strength of the interaction decreases with the size of the organic grouping on the ion.

Such interactions are within the normal experience of chemists in all fields of the subject. In the remainder of the chapter, therefore, we shall confine our attention to more subtle and, in many ways more important, considerations of selectivity which are functions of the size, hydration, and charge of competing counter ions. In this way it is hoped to be able to correlate selectivity with the peculiar physical situation existing within an ion exchanger matrix, which has many of the properties of a concentrated ionic solution under a pressure of several hundred atmospheres.

3-6a. Electroselectivity

It is observed that exchangers with strong acid or strong base functional groups show well-defined selectivities. For polystyrene sulphonic acid (PSSA) exchangers in the range of cross-linking around 10% DVB in dilute aqueous solution, the selectivity series is:

$$Ba^{2+} > Pb^{2+} > Sr^{2+} > Ca^{2+} > Mg^{2+} > Ag^{2+} > Cs^{+} > Rb^{+} > K^{+} > Na^{+} > (H^{+}) > Li^{+}$$

(The position of the hydrogen ion is somewhat less well defined.)

For quaternary ammonium strong base exchangers a similar general series may be observed.

$$\text{citrate}^{2+} > SO_4^{2-} > \text{oxalate}^{2-} > I^- > NO_3^- > CrO_4^{2-} > Br^- > SCN^-$$
$$> Br^- > SCN^- > Cl^- > \text{formate}^- > \text{acetate}^- > F^-$$

In both cases it will be noticed that there is a greater tendency for the exchanger phase to select multivalent counter ions. The electroselectivity term arising from the Donnan membrane potential is a satisfactory explanation for this both qualitatively and quantitatively (see section 3.3 and the Appendix).

It is significant that strongly basic anion exchangers show slight deviation from this electroselectivity trend. In general it is more difficult to rationalise the behaviour of anion exchangers and it is becoming more and more usual to postulate that such exchangers are more prone to ion association between counter ion and fixed ion group, which, if strong, would in some cases overlay the electroselectivity effect.

3-6b. Swelling pressure effects

Considering for a moment only cation exchangers, we see that in the absence of rather specific types of ion binding the exchanger prefers ions primarily according to charge; the electroselectivity term. To examine the selectivities for ions of the same charge we may choose the alkali metal cation series, which, having noble gas electronic configuration, should not enter into specific binding other than possibly ion association. The selectivity series in PSSA resin of moderate cross-linking in dilute solution is $Cs^+ > K^+ > Na^+ > Li^+$, while the relative swelling of the exchanger in the pure ionic forms is in the reverse order, lithium having the largest equivalent swollen volume and caesium the least. Since work must be done in swelling an exchanger bead, the more highly swollen ionic forms will have a contribution to the free energy from the pressure-volume term. Work must be done to swell the exchanger and this may be partly released when the exchanger is converted to a less swollen form.

Although the selectivity sequence for carboxylic acids resins is the reverse[1] of that for sulphonic acid exchangers, so also is the swelling sequence. The swelling pressure sequence is therefore apparently of great importance, especially in cases where no serious bonding occurs between counter ion and fixed charge.

An even more striking example of the way in which selectivity and equivalent volume of swollen resin parallel one another occurs with the dibasic phosphonic acid resins.[1] In acid solutions only the first dissociation is available:

$$\text{—}PO_3H_2 \longrightarrow \text{—}PO_3H^- + H^+$$

This reaction lies almost completely to the right and in this state the exchanger is of the strong acid type, having a selectivity sequence for the alkali metals

$K^+ > Na^+ > Li^+$. If, however, the external solution is alkaline, the second (weaker) acid dissociation is complete:

$$\text{—}PO_3H^- \longrightarrow \text{—}PO_3^{2-} + H^+(OH^-)$$

The anionic grouping is now divalent, a weak acid, and as with carboxylic acid resins, has the reverse selectivity for the alkalies; $Li^+ > Na^+ > K^+$. The equivalent volumes (of the resin forms now) lie in the reverse sequence with the lithium form smallest. Thus whatever the fundamental reason for such sequences in swollen volumes, we must observe that in general an ion exchanger will prefer that ion which will reduce its swollen size and hence its swelling pressure.

3-6c. A more quantitative estimation of the importance of swelling

In this respect, therefore, the most suitable expression for the thermodynamic equilibrium constant is one which will isolate the pressure–volume contribution as a discrete term. [See Appendix, Eqn. (A-10).]

$$\ln K_A{}^B = \ln \frac{\gamma_B{}^{Z_A}}{\gamma_A{}^{Z_B}} + \ln \frac{\bar{\gamma}_A{}^{Z_B}}{\bar{\gamma}_B{}^{Z_A}} + \frac{\pi}{RT}(Z_B \tilde{V}_A - Z_A \tilde{V}_B) \tag{3.12}$$

$\tilde{V}_A$ and $\tilde{V}_B$ are the partial molar volumes of the exchanging ions, π the osmotic or swelling pressure, R the gas constant and T the absolute temperature. The first theory of note on selectivity was that of Gregor which explained the observed selectivities in terms of an elastic resin matrix which, in its inherent resistance to swelling, would prefer the ion with smallest hydrated volume. Gregor's model,[3] if correct, would place most weight upon the pressure term in estimating the sign and magnitude of $K_A{}^B$.

At this stage it is useful to examine Eqn. (3.12) in terms of known data.

(a) If $\tilde{V}_A > \tilde{V}_B$ then $K_A{}^B > 1$; the exchanger prefers the ion with the smaller partial molar volume.

(b) If B were divalent and A monovalent, then $Z_B = 2$ and $Z_A = 1$.

Provided the partial molar volumes differ by less than a factor of two (the usual situation), the ion with larger charge will be selected. This is in accord with the principle of electroselectivity.

(c) If $\tilde{V}_A - \tilde{V}_B$ does not change sign, then $K_A{}^B$ will not either.

(d) If $\tilde{V}_A > \tilde{V}_B$, then as $\bar{X}_B$ increases the resin will contract, since now it contains a smaller proportion of the ions of larger volume. This contraction implies a lower value of π and so $K_A{}^B$ decreases as $\bar{X}_B$ increases. The selectivity for the preferred ion decreases as the proportion of that ion increases in the exchanger phase. This is typical of the behaviour of PSSA resins, as illustrated in Fig. 3.5,[4] which further indicates the influence of cross-linking and thus π upon the selectivity.

Predictions (a), (b) and (d) are qualitatively correct in most cases. It is, however, not uncommon to observe inversions of selectivity as an ion exchanger is progressively converted from one form to another, nor to observe 'crossovers', in which the preferred ion has a lower selectivity in a higher cross-linked exchanger above a certain degree of conversion. Since the swelling pressure in the exchanger is always positive and greater the greater the degree of cross-linking, these observed phenomena can only be explained by assuming that the partial molar volume of the ions may change, or even reverse their relative sizes.

In Gregor's theory,[3] where $\tilde{V}_A$ and $\tilde{V}_B$ are taken as the molar volumes of the hydrated ions, it must then be assumed that in an exchange in which selectivity reverses, the larger ion must be partially or completely stripped of its hydration sheath. Such a process might occur as the pore volumes, in contraction, become too small to allow the ion to maintain its normal hydration sheath. From what is known of the hydration energies of ions it appears unlikely that this plausible explanation is reasonable. Fundamentally it is also a criticism to attempt to use the hydrated ion as a basic species. Hydration is a notoriously difficult concept in chemistry and hydration numbers for ions are not clearly defined, being quite variable according to the type of experiment used in their estimation. The distinction between bound and free water is essentially arbitrary except where the ion actually forms coordinate bonds with them.

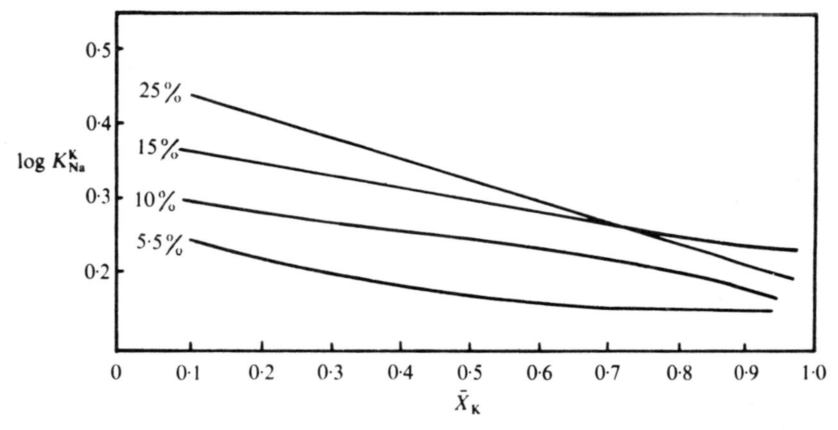

FIG. 3.5. Potassium–sodium selectivity of polystyrenesulphonic acid resins as a function of cross-linking. (Taken from reference 4, with permission)

Glueckauf,[5] neglecting the solvated ion treatment, has applied a more rigorous approach to the problem, considering the ion exchanger as being essentially a concentrated electrolyte solution under pressure, evaluating the pressure–volume and activity terms from water activity data in very low cross-linked resins. His results predict accurately the observed selectivity coefficients for Dowex 50 in hydrogen–alkali metal ion equilibria (Table 3.1).

TABLE 3.1.[a] The Relative Importance of the Activity and Pressure–Volume Terms in Eqn. (3.12). (From Glueckauf, reference 5).

M	Pressure–Volume Term ($\log_{10}$)	Activity Terms ($\log_{10}$)	$K_H{}^M$ (Calculated)	$K_H{}^M$ (Observed)
Li^+	0·00	0·141	1·4	1·3
Na^+	0·01	0·311	1·9	1·8
K^+	0·06	0·452	2·5	2·4

[a] In this table Eqn. (3.12) is expressed in terms of $\log_{10} K_A{}^B$ and not natural logarithms.

In this it is obvious that the pressure–volume term is negligible and that the analogy between concentrated electrolyte and pore solution is a useful one. Purely chemical interactions such as specific matrix interactions cannot be accounted in this theory. These and the fundamental heterogeneity of exchanger structure must give rise to abnormal and essentially unexplained behaviour.

Hogfeldt,[6] in a careful study of the silver–hydrogen ion exchange on Wolfaht KS, has been able to explain the abnormal variations in selectivity with composition of the exchanger only in terms of the existence of a variety of exchanger sites, each with its own typical selectivity pattern. Since all the fixed charges are sulphonate groups, such chemical heterogeneity can only be due to the geometry of these sites, which will be particularly sensitive to local variations in cross-linking.

It therefore becomes apparent that the widely observed relationship between selectivity and swelling is largely an effect rather than a fundamental cause of selectivity behaviour. It is even further eroded by the fact that liquid ion exchangers have similar selectivity patterns to their resin counterparts. Since these are solutions of long chain sulphonates or quaternary ammonium salts dissolved in water-immiscible solvents, they can have no internal swelling pressure.

3-6d. Ion interactions in the exchanger phase

In the previous section the swelling pressure effect upon selectivities was considered.

In the case of the sulphonic acid exchangers in the alkali metal forms, all evidence points to the counter ions being present in the matrix in essentially their normal hydrated forms. In the carboxylic and the fully dissociated phosphonic resins, the very low swelling in the lithium form suggests that the lithium, which is strongly hydrated in aqueous solution, does not retain full hydration in the matrix. Hydrated lithium ion is considered the largest of the alkali metal ions in solution. In exchangers with normal cross-linkings we may discount any possibility of the lithium ion having to shed water of hydration before being able to enter the matrix.

It would seem, therefore, that we must propose some form of association between the lithium counter ion and the carboxyl or $—PO_3^{2-}$. Such association would have the effect of lowering the activity of the resin electrolyte (cation plus fixed anion) and raising the activity of the resin water. This effect must be assumed to decrease as the alkali metal ions increase in atomic number. From Eqn. (3.1) it is obvious that such an increase of water activity in the resin phase must reduce swelling pressure. The least swollen ionic form of the carboxylic group is the hydrogen form where the associated carboxylic group is almost complete.

In order to advance this hypothesis more firmly we must look to evidence from concentrated electrolytes since direct evidence from the resin is not available. Even in aqueous solutions of strong electrolytes there is nothing approaching a quantitative theory at concentrations above 0·2M. Our approach must be qualitative.

The influence of water structure and solvation of ions upon their interaction in concentrated solution will now be considered. Simple inorganic ions may be classified into two types according to as they cause a net orientation of water molecules in their immediate vicinity: (a) order-producing, or (b) order-destroying due to their large size and small charge to surface area ratio. These tend to disrupt the hydrogen-bonded structure of water.[7]

Examples of order-producing ions are:

H^+, Li^+, Mg^{2+}, Be^{2+}, OH^-, F^- and CH_3COO^-

They are usually small ions or have a large charge so that their charge-to-surface area ratio is large. In consequence, these ions bind water molecules in a tightly-knit hydration sheath.

Examples of order-destroying ions are:

Cs^+, Rb^+, K^+, the halide ions other than fluoride, and other large univalent anions such as perchlorate, ClO_4^-

Such ions are not hydrated in the usual sense. They cause a net increase in the disorder of water in their immediate vicinity; that is, they disrupt the existing hydrogen-bonded structure of the water solvent.

Gurney, in particular, has used this approach to rationalise the observed order of the activity coefficients of strong electrolytes. Typical experimental data show that salts of ions of similar ordering character, such as CsCl or LiOH, have lower activity coefficients than those with components of opposite ordering effect, such as CsOH and LiCl. An abnormally low activity coefficient in concentrated solution is taken to imply a net interaction between component ions, which amounts to an attraction. This is explained by Gurney in terms of a lowering of the free energy of the electrolyte due to overlap of ion co-spheres (those areas around the ion in which water molecules are strongly influenced by the ion).

With ions of the opposite ordering character there is a net repulsion when co-spheres overlap in concentrated solution. The large value of the activity coefficient in such cases shows that the free energy of the salt is high. Association of any kind would lower the free energy. This type of ion association is between ions of strong electrolytes in the usual sense of the word. It is also a qualitative phenomenon, since in concentrated solutions we have no theory of ion interactions such as the Debye-Hückel theory to define 'normal' behaviour. The nature of the interactions proposed by Gurney have not been worked out in detail; to this moment, however, their predictions have not been contradicated by experimental fact.

The resin phase of an ion exchanger amounts to a concentrated electrolyte solution and so we shall apply Gurney's arguments to exchange in the manner of Holm.[8] In common strong acid exchangers the sulphuric group is large, singly charged and order-destroying, while in the weak acid ones, the carboxylic group (like acetate) is order-producing. Thus in the sulphonic acid exchangers the tendency to some form of association with the alkali metal ions would be in the order $Cs^+ > K^+ > Na^+ > Li^+$ (as is the order for exchange selectivity). Since the least associated Li^+ is also the most strongly hydrated, the observed swelling order would be Li^+, Na^+, K^+, Cs^+, as is indeed observed.

On the other hand, the carboxyl group being order-producing would have greatest tendency to associate with the order producing Li^+ ion and least with Cs^+, again paralleled by the observed selectivity order. In ion pairing or ion association some charge neutralisation will occur due to the close proximity of the ions of the pair. It is not surprising therefore that the Li^+ form would also be the least swollen form.

The swelling observed is of course a function of the swelling pressure, which in turn is given by Eqn. (3.1). Since the more concentrated the electrolyte the lower the activity of water, the left hand side of this equation is positive and so the matrix is under a positive internal pressure. If, on the other hand, there is a net association of electrolyte, there is an increase in the water activity over the unassociated case. A similar effect in the resin phase would cause an increase in $\bar{a}_w$ and so a decrease in swelling pressure. On this basis the ionic form with most association in the resin phase would be least swollen.

Whitney and Diamond[9] have looked at this problem of selectivity in a slightly different way, not at all contradicting the theories of Gurney and Holm. In the system

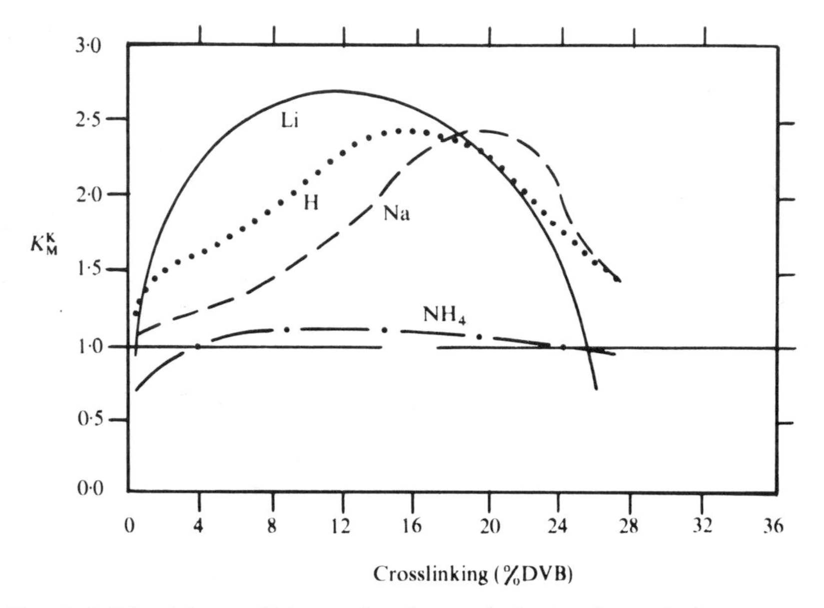

FIG. 3.6. Selectivity coefficients of various univalent cations relative to potassium as a function of the degree of cross-linking. The exchanger used was polystyrenesulphonic acid resin. (Taken from reference 10, with permission)

exchanger and dilute solution, there are two separate phases. The resin phase, which may be from 4 to 7 molal in internal electrolyte concentration, has a very much smaller ratio of water molecules to ions than the dilute external phase. Under these conditions counter-ion solvation may be less complete in resin than in solution. The sulphonate ion, being less polarisable than the water molecule, is a less competent 'solvating' agent. On this model we may explain the selectivity of PSSA resins for the heavy alkali metal ions. In an exchange involving, say, lithium and caesium ions, the lithium ion will favour the solution phase, where it can obtain maximum solvation and leave the caesium ion, which is essentially unsolvated in water, to enter the resin phase with its poorer solvating tendency.

Fixed carboxylate anions in the resin phase would constitute a much more powerful solvating species than the sulphonate and thus the selectivities for the alkali metal ions on such weak acid exchangers is less pronounced or even reversed.

Anion selectivities on strong base exchangers are on the whole less swollen than sulphonic resins of comparable cross-linking. Low swelling implies a large water activity and consequently a low resin electrolyte activity, suggesting that ion association may be a dominant feature in the selectivity sequences on these resins.

3-6e. The effect of cross-linking

This discussion will be confined to sulphonic resin exchangers, which are the only ones for which there appears sufficient data. Gregor and Bregman[10] have measured selectivity coefficients on such resins with varying degrees of cross-linking, taken as being proportional to the DVB percentage composition of the polymer. Their results are shown in Fig. 3.6. At 8% DVB (comparable to general purpose resins) the order of resin selectivities are $K^+ > NH_4^+ > Na^+ > H^+ > Li^+$. As the degree of cross-linking decreases, the selectivity coefficients approach unity except for the ammonium ion, which appears to have some residual selectivity over potassium even at zero cross-linking; this is possibly due to ion association. The effect of increasing cross-linking is to increase selectivity, reaching a maximum around 8–12% DVB.

Although selectivities generally approach unity at zero cross-linking it is not possible to ascribe all selectivity to the pressure effects attendant upon cross-linking. Cross-linking also increases internal concentrations, bringing the resin ions into closer contact, and so has a major effect upon activity coefficients. This as we have seen is the less obvious, but more important source, of exchanger selectivity.

REFERENCES

1. J. I. Bregman, *Ann. N.Y. Acad. Sci.* **57,** 125 (1954).
2. K. A. Kraus, H. O. Phillips, T. A. Carlson and J. S. Johnson, *Intern. Conf. Peaceful Uses Atomic Energy, Geneva* **28,** 3 (1958).
3. H. P. Gregor, *J. Amer. Chem. Soc.* **70,** 1293 (1948); **73,** 642 (1951).
4. D. Reichenberg and D. J. McCauley, *J. Chem. Soc.* **1955**, 2731.
5. E. Glueckauf, *Proc. Roy. Soc. (London) Ser. A.* **214,** 207 (1952).
6. E. Hogfeldt, *Arkiv. Kemi* **13,** 491 (1959); **7,** 561 (1954).
7. R. W. Gurney, *Ionic Processes in Solution*, McGraw-Hill, New York, 1953.
8. L. W. Holm, *Arkiv Kemi* **10,** 461 (1956).
9. D. C. Whitney and R. M. Diamond, *J. Inorg. Nucl. Chem.* **27,** 219 (1965), and R. M. Diamond, *Inorg. Chem.* **2,** 1284 (1963).
10. H. P. Gregor and J. I. Bregman, *J. Colloid Sci.* **6,** 323 (1951).

4 Kinetics of ion exchange

Those familiar with simple homogeneous chemical reactions are acquainted with the common expressions in which the rate, R, of the reaction (4.1) may be represented by Eqn. (4.2).

$$A + B \rightarrow \text{products} \tag{4.1}$$

$$R = k \cdot (A)^x \cdot (B)^y \tag{4.2}$$

where k is the rate constant and (A) and (B) are the concentrations of reacting species. The powers x and y are the orders of the reaction with respect to A and B respectively.

The ion exchange process is, however, quite different in many respects from such simple chemical reactions. In the first case it concerns a reaction which involves the transport of ions from solution into exchanger and vice versa and is therefore heterogeneous. For dilute solutions, there is effectively no electrolyte penetration into the exchanger and hence the co-ion has no part to play in the overall reaction mechanism.

$$X^-A_{soln.}^+ + (B^+—R^-) \rightarrow X^-B_{soln.}^+ + (A^+—R^-) \tag{4.3}$$

In an ion exchange such as reaction (4.3), an electrolyte containing counter ion A^+ is exchanging with a matrix wholly in the B^+ form.

Since no co-ion effectively enters the exchanger, the process must be one which at all times preserves the electroneutrality of each phase. The number of charges carried into the exchanger by A ions entering must equal the number of B ion charges leaving in the same time. This coupling of flows of the entering and leaving ions does much to simplify the treatment of exchange kinetics.

A typical experimental situation is one in which a sample of the exchanger particle or beads in the B form is added to a stirred solution containing counter ion A. Since the rate of a chemical reaction is determined by its slowest kinetic step, it is well to consider these steps in turn. Even in a well-stirred solution there remains a layer of unstirred solution close to any solid surface which may be immersed in that solution. In particular around an exchanger bead, we will have such a layer which under good stirring conditions will be of the order of 10^{-3} cm thick. Being unstirred, the only manner in which the ion may cross will be by diffusion; the film diffusion step.

On entering the exchanger, the ion must find its way through the pores of the matrix to a fixed site occupied by a B ion; again a process of diffusion. At the site the B ion will exchange for A in what is a chemical reaction step:

$$A + \bar{B}\bar{R} \rightarrow B + \bar{A}\bar{R} \tag{4.4}$$

The B ion by the reverse diffusion steps must then find its way to the outside stirred solution.

There are therefore three distinct kinetic processes:

1. Film diffusion in the unstirred layer of solution close to the exchanger particle.
2. Particle diffusion of ions in the matrix.
3. Chemical exchange reaction.

Of all the exchange kinetics which have appeared in the literature to date, none have been shown to be controlled by the chemical exchange step. This is not surprising in the case of the strong acid and strong base exchangers where the fixed ionic group forms no chemical bond with common counter ions, but is somewhat remarkable when the ion exchange site is a complexing ligand and the counter ion one which is well known as being able to form stable and kinetically inert complexes. In an attempt to obtain chemically controlled ion exchange, rates of exchange involving cobaltous ion with Dowex A-1 resin have been studied. This exchanger contains as the functional group iminodiacetate complexing fixed groups (Fig. 3.4). Even in this case, however, the kinetically slow step was found to be particle diffusion.[1]

In practice, therefore, we will confine our discussion of ion exchange kinetics to those controlled by diffusion steps in the film or in the exchanger matrix: film diffusion and particle diffusion, respectively.

Consideration of the diffusion problem is much simplified by the fact that the flows or fluxes of exchanging ions in the exchanger are electrically coupled. Regardless of the inherent mobility of the ions concerned they will migrate with the same diffusion coefficient. This will become obvious when we consider what will happen if, say, B^+ migrates more quickly than A^+ in the exchanger. B^+ ions will tend to leave the exchanger phase in greater numbers than A^+ ions will enter. The net effect would be to leave some fixed charge (negative) uncompensated on the matrix chains, so causing a separation of charge and an electrical potential negative in the matrix and positive in the solution. The effect of this potential will be to speed up the entering A ion and correspondingly slow the leaving B ion until their fluxes are equal and opposite at any point in the exchanger. The electrical coupling is a consequence of the requirement for electrical neutrality within the phase.

4-1. ISOTOPIC EXCHANGE

This is the simplest experiment from a theoretical view. The system exchanger–solution is in chemical equilibrium except for isotopic distribution, exchanging ions being merely different isotopes of the same counter ions or co-ion species. This being so there are no gradients of electrical potential in solution or exchanger and no gradients of activity coefficients, and so the flow of isotope, J_i, at any point in the exchanger is given by Fick's law (compare with section 4.3).

$$J_i = -\bar{D}_i \operatorname{grad} \bar{C}_i \tag{4.5}$$

where grad $\bar{C}_i$ is the gradient of concentration of i at that point, $\bar{D}_i$ the isotopic diffusion coefficient, usually in cm^2/sec and the units of flow are g ion/cm^2 sec.

4-2. PARTICLE DIFFUSION

The mathematical problem of determining the rate at which exchange occurs is quite analogous to the rate of cooling of a hot sphere placed in a well-stirred thermostat bath. When the beads are placed in contact with solution a very rapid exchange occurs as isotope is removed from the outer surface of the beads and concentration gradients are extremely large, being confined to surface layers of the exchanger, Eqn. (4.5). As the exchange proceeds, exchange sites are closer to the centres of the beads and the concentration gradients for isotopes are reduced. The rate of exchange is therefore characterised by an initially rapid step followed by progressively slower rates as the exchange proceeds. When a sample of exchanger in the i-form is placed in a well-stirred solution of large volume (infinite bath conditions) initially devoid of isotope i, the rate of the exchange reaction is given by Eqn. (4.6),[2]

$$U_{(t)} = [1 - \exp(\bar{D}t^2/r_0^2)]^{1/2} \tag{4.6}$$

$U_{(t)}$ is the fractional attainment of equilibrium, defined as

$$U_{(t)} = \frac{\text{amount of } i \text{ exchanged in time } t}{\text{amount of } i \text{ exchanged at time infinity (equilibrium)}}$$

The half time of the exchange, $t_{\frac{1}{2}}$, may be obtained by substituting $U_{(t)} = 0{\cdot}5$ in Eqn. (4.6), giving Eqn. (4.7):

$$t_{\frac{1}{2}} = 0{\cdot}030 r_0^2/\bar{D} \tag{4.7}$$

In both equations t is measured in seconds, $\bar{D}$ in cm^2/sec and r_0, the bead radius, in cm. A plot of $U_{(t)}$ versus t is characterised by a very rapid initial exchange followed by subsequent falling off as concentration gradients diminish within the bead. The rate of exchange is proportional to the diffusion coefficient in the bead and inversely proportional to the square root of the bead radius. Doubling the size of the beads would reduce the rate by a factor of four. The values of self-diffusion coefficients and their interpretation are discussed at greater length in section 5.3.

4-3. FILM DIFFUSION

In contrast, film diffusion controls the rate of exchange when the slow step is diffusion across the unstirred solution layer at the surface of the exchanger bead. Concentration gradients occur only in the solution phase. For infinite bath conditions, in which isotope i remains essentially zero in the bulk solution throughout the exchange, the expression for $U_{(t)}$ becomes Eqn. (4.8).[3]

$$U_{(t)} = 1 - \exp(-3DCt/r_0\delta\bar{C}) \tag{4.8}$$

C is the total solution concentration, barred for exchanger; D the solution diffusion coefficient; δ the thickness of the unstirred solution layer. Again the half-time of reaction, $t_{\frac{1}{2}}$, is obtained by substitution, $U_{(t)} = 0{\cdot}5$.

$$t_{\frac{1}{2}} = 0{\cdot}23 r_0\delta\bar{C}/DC \tag{4.9}$$

The rate of exchange is proportional to the diffusion coefficient and the concentration of the isotope in the solution. It is inversely proportional to the bead radius, film thickness and total concentration of the species in the exchanger. For a typical ion exchanger, capacity 1 mequiv/ml in a well-stirred solution, $\delta \approx 1 \times 10^{-3}$ cm; $r_0 = 0{\cdot}1$ cm, and $D = 1 \times 10^{-5}$, exchanging counter ions with a solution 10^{-3} molar, $t_{\frac{1}{2}}$ is approximately 40 min. The general form of the rate curve is shown in Fig. 4.1.

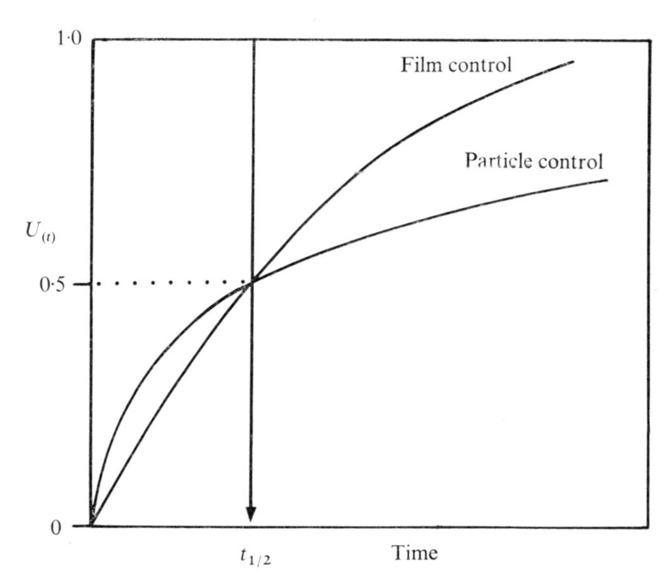

FIG. 4.1. Comparison of $U_{(t)}$ versus t plots for film and particle diffusion control under conditions in which each mechanism would give the same half-time, $t_{\frac{1}{2}}$, of exchange.

4-4. ION EXCHANGE

A quantitative treatment of ion exchange is much more complicated than isotopic exchange, which occurs in a system essentially in equilibrium. The exchanging ions A and B, Eqn. (4.4), are in principle quite different in valency and in mobility – complete exchange may well involve considerable water transport and consequent swelling changes. Within the exchanger, ions are subject to electrical and diffusional forces during the exchange process and the total flow of an ion, i, is considered to be made up of the sum of the diffusional and electrical flows (Eqn. 4.10).

$$J_i = J_{i(\text{diff})} + J_{i(\text{el})} = -D_i \left(\text{grad } C_i + \frac{z_i C_i F}{RT} \text{grad } \varphi\right) \tag{4.10}$$

Equation 4.10 is known as the Nernst–Planck equation (see section 5.2). D_i, C_i and J_i have their usual meanings. F, R, and T are the Faraday, the gas constant and the absolute temperature respectively, while grad φ is the gradient of the electrical potential in the medium. This equation may be applied with equal validity to solution or exchanger. Under the restriction that $\bar{D}_A$ and $\bar{D}_B$, the self-diffusion coefficients, are constant and that co-ion uptake and swelling changes are negligible it is possible to make a quantitative analysis of ion exchange kinetics. There are two basic restrictions in this process of ion exchange. The first is that there will be electroneutrality in the bead at all times so that Eqn. (4.11) holds.

$$z_A\bar{C}_A + z_B\bar{C}_B = \text{concentration of fixed charges (constant)} \qquad (4.11)$$

Differentiating,

$$z_A \text{ grad } \bar{C}_A + z_B \text{ grad } \bar{C}_B = 0$$

For ions of like charge the concentration gradients of A and B at any point and at any time during the exchange are equal and opposite in sign.

The second condition is that there will be no net transfer of charge and hence no electric current. The number of equivalents of charge entering and leaving the exchanger are equal [Eqn. (4.12)].

$$z_AJ_A + z_BJ_B = 0 \qquad (4.12)$$

In consequence, for ions of equal charge, $J_A = -J_B$ and grad $\bar{C}_A = -$grad $\bar{C}_B$, although in principle $\bar{D}_A \neq \bar{D}_B$. From substitution for J_A and J_B in the Nernst–Planck equation [Eqn. (4.10)], it becomes obvious that it is the size and the direction of the electrical potential gradient that ensures the coupling of the flows of A and B.

Equation (4.10) applied to each ion may be combined with the limiting conditions of Eqns. (4.11) and (4.12). Elimination of electrical potential terms from these gives an equation for J_A in terms of grad $\bar{C}_A$.

$$J_A = -\bar{D}_{AB} \cdot \text{grad } \bar{C}_A \qquad (4.13)$$

This has the form of Fick's law where $\bar{D}_{AB}$ is the mutual diffusion coefficient defined as in Eqn. (4.14).

$$\bar{D}_{AB} = \frac{\bar{D}_A\bar{D}_B(z_A{}^2\bar{C}_A + z_B{}^2\bar{C}_B)}{(z_A{}^2\bar{C}_A\bar{D}_A + z_B{}^2\bar{C}_B\bar{D}_B} \qquad (4.14)$$

The mutual diffusion coefficient is therefore not a constant since it is a function of the varying concentrations of A and B.

When $t = 0$,
$z_A\bar{C}_A =$ capacity, $z_B\bar{C}_B = 0$ and $\bar{D}_{AB} = \bar{D}_B$

When $t = \infty$,
$z_A\bar{C}_A = 0$, $z_B\bar{C}_B =$ capacity and $\bar{D}_{AB} = \bar{D}_A$

The ion present as the *minor* component dominates the interdiffusion, at first sight a remarkable conclusion. In the terms of Eqn. (4.10), if A is a minor component, $\bar{C}_A$ will be of the order of zero and so the electrical contribution to the flow (which is proportional to $\bar{C}_A$) will also be small, approaching zero. For the minor component, therefore, $J_A \approx \bar{D}_A$ grad $\bar{C}_A$, simple diffusion with concentration gradient being the major contribution to the total flow. Helfferich and Plesset,[4] who originated this approach, have calculated interdiffusion coefficients as functions of ionic composition of the exchanger (Fig. 4.2). Isotopic exchange is therefore a special case of this treatment under conditions where $\bar{D}_A$ and $\bar{D}_B$ are equal.

4-4a. Rates of forward and reverse exchanges

Helfferich, calculating an explicit equation for the degree of conversion $U_{(t)}$ as a function of time, has illustrated that the rate of forward and reverse exchanges may differ quite markedly when the ratio $\bar{D}_A/\bar{D}_B$ is large. In hydrogen–sodium exchanges this ratio is around seven and the half-times differ by a factor of two. The exchange is faster when the faster ion (hydrogen ion, in this case) is initially on the exchanger.

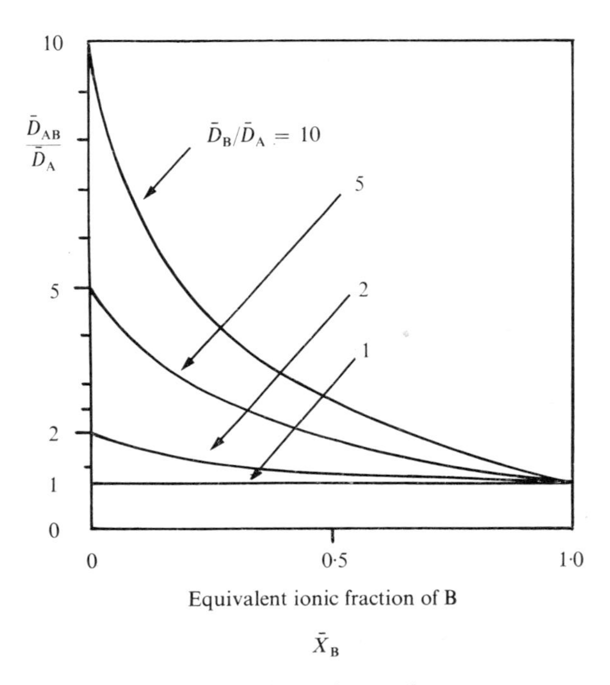

FIG. 4.2. Interdiffusion coefficient in an ion exchanger as a function of ionic composition in the exchanger phase. $\bar{D}_{AB}$ is the mutual diffusion coefficient in the exchange reaction $\bar{A} + B \rightarrow \bar{B} + A$. $\bar{D}_A$ and $\bar{D}_B$ are the tracer diffusion coefficients of A and B in the pure A and pure B-forms respectively. (Taken from reference 4, with permission)

This is a consequence of the dependence of the mutual diffusion coefficient upon composition.

Upon the initial immersion of the H-form in a solution of sodium ions the outer layers of the exchanger beads become almost totally populated with sodium ions, the concentration of hydrogen ions being minimal. Under these conditions the mutual diffusion coefficient is of the order of $\bar{D}_H$. A fast exchange is initiated and the outer shells are quickly converted, although as the exchange proceeds the rate reduces progressively until $\bar{D}_{AB} \approx \bar{D}_{Na}$ in the latter stages. In the reverse direction the opposite situation holds and the mutual diffusion coefficient increases as the conversion proceeds.

4-4b. SORPTION AND DESORPTION OF ELECTROLYTES

The requirement of electroneutrality within the exchanger constrains the rate of uptake of salt counter ion and co-ion to be coupled. Defining a salt AX with counter ion A and co-ion X, the restriction of zero electric current requires the relationship (4.15) (see section 5.5).

$$(z_A J_A + z_X J_X) \cdot F = I = 0 \tag{4.15}$$

For a 1:1 electrolyte,

$$z_A = -z_X = 1 \text{ so that } J_A = J_X$$

The cation and anion must diffuse at the same rate.

The second restriction is that in the process of salt diffusion, electroneutrality is preserved.

$$z_A \bar{C}_A + z_X \bar{C}_X = \text{concentration of fixed charges} \tag{4.16}$$

With these two restrictions and the Nernst–Planck Eqn. (4.10) for each ion, the mutual diffusion coefficient, $\bar{D}_{AX}$, for the salt may be calculated, in the manner of section 4.3. The final expression for $\bar{D}_{AX}$ is Eqn. (4.17).

$$\bar{D}_{AX} = \frac{D_A D_X (z_A{}^2 \bar{C}_A + z_X{}^2 \bar{C}_X)}{z_A{}^2 \bar{C}_A \bar{D}_A + z_X{}^2 \bar{C}_X \bar{D}_X} \tag{4.17}$$

By the Donnan exclusion principle (section 3.3) $\bar{C}_X$ is very much less than $\bar{C}_A$ under normal conditions. To a close approximation we may therefore neglect $\bar{C}_X$ in Eqn. (4.17), so that $\bar{D}_{AX} \approx \bar{D}_X$. The salt diffusion coefficient is equal to the self-diffusion coefficient of the co-ion (regardless of the counter ion mobility) and is effectively constant. It may, therefore, be treated by the laws of isotopic exchange (section 4.2). This model neglects the effects of swelling and shrinking which will occur to some degree as an exchanger is placed in a lesser or more concentrated solution.

4-5. EXPERIMENTAL DISTINCTION BETWEEN FILM AND PARTICLE DIFFUSION-CONTROLLED EXCHANGES

The observed rate of exchange may be either particle, film, or mixed particle–film diffusion controlled. Particle diffusion control is caused by a slow diffusion rate in the exchanger. Concentration gradients are confined to the exchanger and are absent in the unstirred solution films. In contrast film diffusion control is caused by a slow diffusion step in the unstirred film and no concentration gradients exist in the particle. The interruption test[5] is the best experimental test for distinguishing particle and particle–film from film control. During a kinetic experiment the exchanger beads are removed and quickly separated from adhering solution. After a brief period they are re-immersed in their solution and the experiment continued. If the control is due to particle diffusion, the rate immediately on re-immersion is enhanced (Fig. 4.3). During the interruption concentration gradients disappear in the particle and on re-immersion are much greater than before at the surface. If control is due to film diffusion, no concentration gradients exist in the exchanger and so the interruption has no effect. Mixed particle–film diffusion control will of course be affected in the same way as pure particle control.

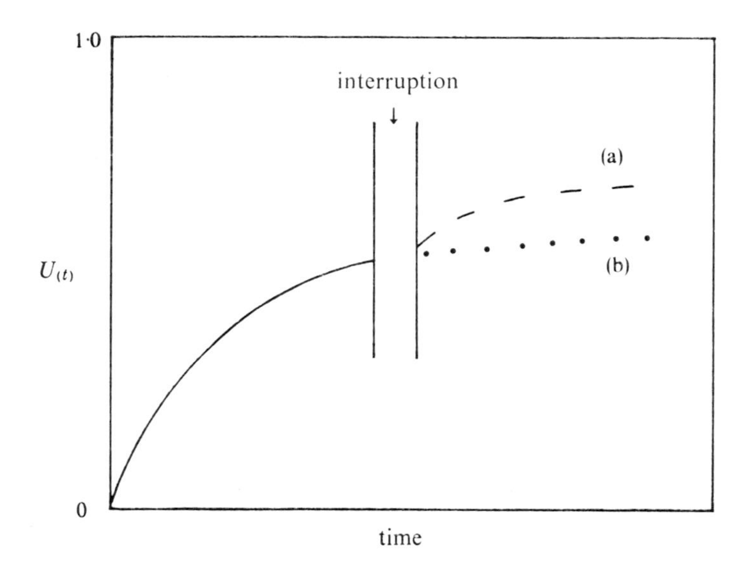

FIG. 4.3. The effect of an interruption test on observed kinetics. (a) Enhanced rate after interruption indicates that the rate controlling step is influenced by concentration gradients in the bead–particle or mixed particle/film diffusion control. (b) If the rate remains unaffected before and after interruption film diffusion control is rate determining.

4-6. THEORETICAL PREDICTIONS OF THE RATE CONTROLLING PROCESS

The nature of the rate controlling step may be predicted[6] from the value of the function

$$\frac{X\bar{D}\delta}{CDr_0}(5 + 2\alpha_B{}^A) = \phi$$

X is the concentration of fixed charges; C, the concentration of solution in equivalents/litre; $\bar{D}$ and D the interdiffusion coefficients in exchanger and in solution, respectively; r_0, the bead radius in cm; δ the film thickness (cm), and $\alpha_B{}^A$ the separation factor, defined as $\bar{C}_A C_B/\bar{C}_B C_A$, where all concentrations are in mole/litre of the medium involved.

If ϕ is very much less than unity the rate is particle diffusion controlled. If very much greater than unity the control is by film diffusion. These criteria are rough but useful for the reaction:

$$B + \bar{A} \rightarrow \bar{B} + A$$

Particle diffusion is favoured by *small:* capacity, X; exchanger diffusion coefficient (high cross-linking); separation factor for the ion initially held by the exchanger; and by *large:* solution concentration, C; solution diffusion coefficient, D; radius, r_0.

In particular, a given ion exchange performed on a given ion exchange sample will have constant X, $\bar{D}$, D, δ, and r_0 and so will change from particle to film control as the solution concentration falls.

REFERENCES

1. A. Schwartz, J. A. Marinsky and K. S. Spiegler, *J. Phys. Chem.* **68,** 918 (1964).
2. T. Vermulen, *Ind. Eng. Chem.* **45,** 1664 (1953).
3. G. E. Boyd, A. W. Adamson and L. S. Mayers, *J. Amer. Chem. Soc.* **69,** 2836 (1947).
4. F. Helfferich and M. S. Plesset, *J. Chem. Phys.* **28,** 418 (1958).
5. T. R. E. Kressman and J. A. Kitchener, *Disc. Faraday Soc.* **7,** 90 (1949).
6. F. Helfferich, *Ion Exchange*, McGraw-Hill, New York, 1962.

FURTHER READING

F. Helfferich, reference 6 and *Ion Exchange Kinetics*, Chapter 2 in *Ion Exchange* (Ed. J. A. Marinsky), Marcel Dekker, New York, 1966.

5
Exchanger membranes

Ion exchange membranes are sheets, ribbons or rods of ion exchange material capable of separating two solutions. Their interest lies in an ability to control the flow of mobile species from one solution to another. In recent years with the development of stable membranes they have become of great interest in industry, with applications in desalination and fuel cell technology, while for purely scientific work have provided the most useful medium of all for the study and elucidation of the ion exchange process. In biology the scientist is faced with systems of enormous complexity in which cell growth, waste disposal and nerve signals are all controlled by a variety of sophisticated biological membranes. The structure and function of these constitutes one of the greatest current challenges in science. There, too, combinations of simple ion exchange membranes are providing the first crude analogues of biological systems and the only well-defined theory of membrane processes.

This chapter will therefore be devoted to a discussion of the transport properties of membranes and the 'forces' which drive an ion or molecule from one environment to another. The dominant tendency of any chemical system is to reach chemical and mechanical equilibrium.

5-1. FLOWS AND FORCES

If two solutions of unequal concentration are placed in contact, a series of diffusion and transport processes will begin which will end only when equilibrium is achieved. Thermodynamics defines the equilibrium condition as one in which the chemical potentials of all species in a system are constant with time and correspond to a minimum in the free energy of the system.

If two solutions of the same chemical composition but different concentration are placed in contact, each component will have different chemical potentials from one solution to another. For simplicity, let us consider only component A, which has chemical potential μ_A' in solution 1 and μ_A'' in solution 2. In the region of contact there will develop a gradient of chemical potential (Fig. 5.1). The flow of A will be from high chemical potential to low; as with water under mechanical potential the flow is 'downhill'. Under the simplest conditions, in which A is without interaction with other components and μ_A' and μ_A'' are constant, there will be a steady flow of A between the two solutions which is proportional to the negative gradient of chemical potential in the interfacial layer, Eqns. (5.1a) and (5.1b).

$$J_A = L\,.\,(-\mathrm{d}\mu_A/\mathrm{d}x) \tag{5.1a}$$

$$-\mathrm{d}\mu_A/\mathrm{d}x = R\,.\,J_A \tag{5.1b}$$

These two equations are simply alternative representations in which J_A is the flux or flow of A in the x direction in moles per unit area per unit time, $-\mathrm{d}\mu_A/\mathrm{d}x$ the thermodynamic force and L and R the permeability and resistance coefficients respectively.

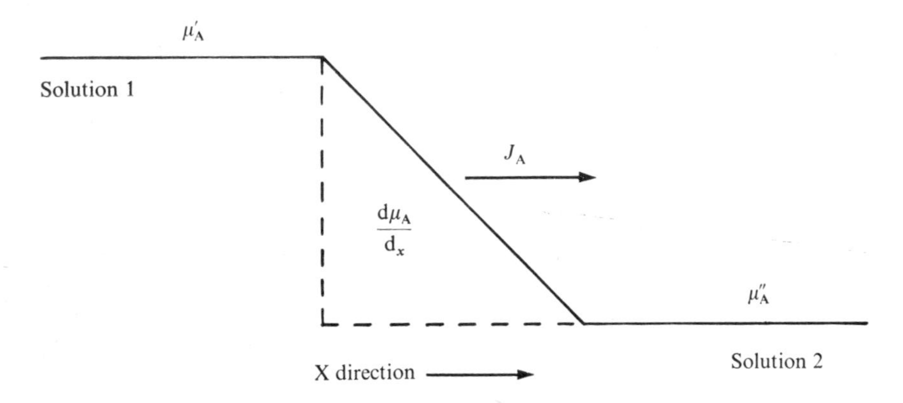

FIG. 5.1. The relationship between the flow, J_A, of a chemical species, A, and the gradient of chemical potential, $\mathrm{d}\mu_A/\mathrm{d}x$, causing that flow. (In this simple case it is assumed that there is only one mobile species)

Since the equations are merely alternative representations of the same equation, R in this case is the reciprocal of L.

These are the primary equations of non-equilibrium thermodynamics. The coefficient L has the same importance as a diffusion coefficient, being large when the medium is easily permeable to A and small when transport is difficult.

Accepting for the moment this simple concept of the relationship between flows and forces, it is useful to consider further the separate components of the chemical potential. For an isothermal system, the chemical potential of the species A may be written as:

$$\tilde{\mu}_A = \mu_A{}^0 + RT \ln a_A + z_A F\varphi + (P - P^0)\bar{V}_A \tag{5.2a}$$

The terms on the right hand side of Eqn. (5.2a) represent all the important contributions to the chemical potential of the species – they are discussed in some detail in the Appendix.

The thermodynamic force, X, is therefore given by

$$X = -\mathrm{d}\tilde{\mu}/\mathrm{d}x = -RT\,\mathrm{d}\,(\ln a_A)/\mathrm{d}x - z_A F\,\mathrm{d}\varphi/\mathrm{d}x - \bar{V}_A\,\mathrm{d}P/\mathrm{d}x \tag{5.2b}$$

The total driving force is made up of contributions from activity, electrical and pressure gradients; some or all of which may be present under experimental conditions.

At this stage it is useful to consider two simple examples. (a) Fick's law of diffusion; this is primarily an experimental law which states that the flow of a species in diffusion (J) is proportional to the negative gradient of concentration, dc/dx.

$$J = -D\, dc/dx \tag{5.3}$$

The constant of proportionality, D, is the diffusion coefficient. If we combine Eqns. (5.1a) and (5.2b), under the conditions that electrical and pressure gradients are absent and allow that concentration, c, may be taken as equal to the activity – to a first approximation – we obtain:

$$J = -L\, RT\, d\,(\ln c)/dx = -\left(\frac{L\, RT}{c}\right) dc/dx \tag{5.4}$$

A comparison of Eqns. (5.3) and (5.4) shows that they have identical form and this constitutes a proof of Fick's law. (b) Ohm's law as a particular case of Eqn. (5.1). In a system in which neither activity nor pressure gradients exist and only electrical force is applied a combination of Eqns. (5.1a) and (5.2) yields Eqn. (5.5).

$$J_A = -L z_A F\, d\varphi/dx \tag{5.5}$$

The flow of A is proportional to the negative gradient of electrical field applied. If we consider J_A to be measured in moles/cm²/sec and the Faraday in coulombs/equivalent, then the current constituted by this flow of A is given by Eqn. (5.6).

$$I = J_A z_A F = -L z_A^2 F^2\, d\varphi/dx \tag{5.6}$$

(*cf.* Section 5.5a)

I is the current density in A/cm²/sec, $d\varphi/dx$ the applied voltage gradient in volts/cm and so $L z_A^2 F^2$ is the specific conductivity of the medium. Such a result is of course a proof of Ohm's law under the specific restrictions that A is a charged species, is mobile, and is the only mobile species in the system: exactly the conditions of an electron in a metal. These two simple examples are taken to show in the simplest possible way the power of the non-equilibrium thermodynamic approach. It shows clearly that a chemical species will flow under a thermodynamic force which is the sum of minor forces caused by activity (concentration), electrical, and pressure gradients.

In the introduction given above the number of mobile species has been restricted to one only for the sake of simplicity. In electrolyte solution and in ion exchangers there are many mobile species; cations, anions, solvent and non-electrolyte. Each will be acted upon by its own appropriate thermodynamic force. There will, however, be cross effects in which the total flow of each species is affected by its interactions with all the others. A very powerful example occurs when we consider the effect of electric force upon a permselective ion exchanger in equilibrium with two identical salt solutions. Electric current will flow in the membrane by migration of the counter ion and no force will be applied directly to the pore water, since it has no charge. There will be a migration of water notwithstanding; this is due to electro-osmosis. The migration is in the same direction as counter ion flow and is due to the electrical interaction between counter ion and water dipole (Fig. 5.2). A force directly on an ion will cause a concurrent flow of water as a secondary effect. Such effects involve flow equations of the type:

$$J_1 = L_{11}X_1 + L_{12}X_2$$
$$J_2 = L_{21}X_1 + L_{22}X_2 \qquad (5.7a)$$

These linear equations allow cross effects, since if the cross coefficients L_{12} and L_{21} are non-zero, a force on 1, X_1, will cause a flow of 2 even although X_2 (the direct

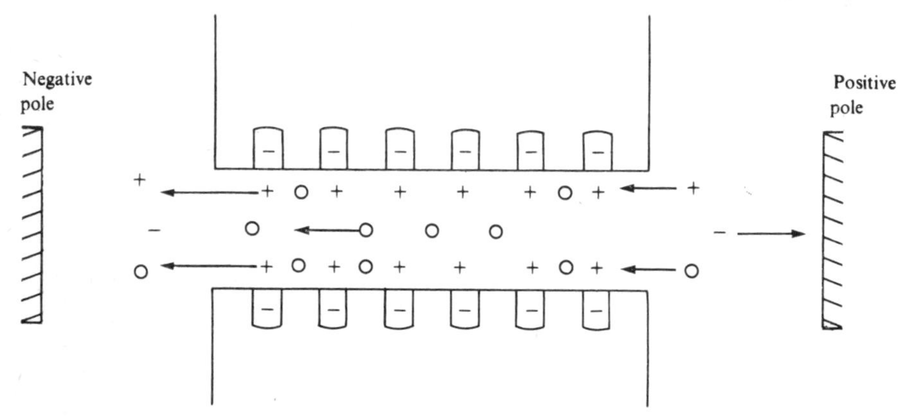

FIG. 5.2. Electro-osmosis. A diagrammatic representation of electro-osmotic flow of water in a cation exchanger pore. Counter ion, co-ion and water are given the symbols +, −, and o respectively

force on 2), is zero. The reverse sequence is also true. Under these conditions the flow of 1 and 2 are said to be coupled. As before there is an equally acceptable alternative representation in terms of frictional coefficients i.e. Eqn. (5.7b).

$$X_1 = R_{11}J_1 + R_{12}J_2$$
$$X_2 = R_{21}J_1 + R_{22}J_2 \qquad (5.7b)$$

In the general case where there are n components in a system the flow of any one, i, is affected by the forces on all the others so that:

$$J_i = \sum_{k=1}^{n} L_{ik}X_k \quad \text{or} \quad X_i = \sum_{k=1}^{n} R_{ik}J_k \qquad (5.8)$$

These are the phenomenological equations of non-equilibrium thermodynamics. They apply only under rather restricted conditions: when the system is close to equilibrium and there is a steady state of flow. Under these conditions and with a proper definition of forces and flows the Onsager reciprocal relations hold. The reciprocal relations state that in the phenomenological equations $L_{ik} = L_{ki}$ (alternatively that $R_{ik} = R_{ki}$). They are of the greatest importance since they reduce the number of coefficients required to define the system. The cross coefficients are due to secondary interactions and are therefore (usually) smaller than the direct coefficients, obeying the inequality (5.9) (and similarly for the resistance coefficients).

$$L_{ii} \,.\, L_{kk} \geqslant L_{ik}^2 \qquad (5.9)$$

It is intuitively obvious that a given force in direct application will cause a greater flow than when applied to a coupled species and transmitted by interactions of the type mentioned in electro-osmosis.

It is useful to compare this approach with the Nernst–Planck equation of Chapter 4. It is then immediately obvious that the Nernst–Planck equation is obtained from Eqns. (5.2) and (5.8), simply by ignoring cross coefficients. In kinetics this is justified, since the experimental results are of a lower order of accuracy than in membrane studies and the deviations from equilibrium in the kinetic process may be large and beyond the rigorous application of this more sophisticated theory.

This rather brief diversion into non-equilibrium thermodynamics, although conceptual, provides a useful model for the interpretation of membrane phenomena.

5-2. SELF OR ISOTOPIC DIFFUSION

The simplest case of diffusion is one in which a membrane is in chemical equilibrium with its bathing solutions, except that on one side a mobile species is isotopically

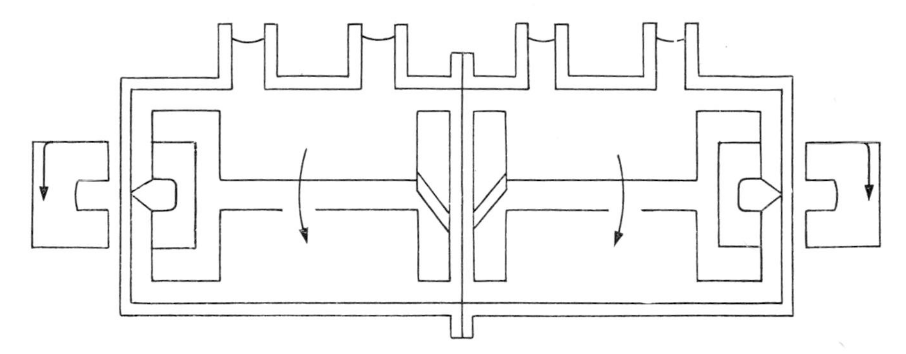

FIG. 5.3. Isotope diffusion cell

labelled. The apparatus consists of two half cells separated by the membrane and provided with stirrers, as shown in Fig. 5.3. On addition of isotope i, there is an initial period (known as the dead time) during which it penetrates the membrane no activity is detectable in the second solution. Thereafter a plot of total radioactivity in the second solution is initially curved and finally linear, indicating the establishment of a steady state of diffusion flow (Fig. 5.3a). In this state the concentration profiles of isotope are stationary across the membrane and the number of isotopes leaving the first equals those reaching the second solution in a given time interval. Since it is quite impossible to efficiently stir any solution right up to a solid interface, there will be two unstirred layers at the membrane surfaces in which isotopic concentrations are not those of the bulk solutions and in which concentration gradients will exist. Since the two solutions are chemically identical, apart from the presence of isotope, there can be no gradients of activity coefficients, electrical potential or pressure across the system which can contribute to the thermodynamic force, X_i, causing diffusion. From Eqn. (5.2)

$$X_i = -\mathrm{d}\mu_i/\mathrm{d}x = -RT\,\mathrm{d}(\ln c_i)/\mathrm{d}x = -\frac{RT}{c_i}\,.\,\mathrm{d}c_i/\mathrm{d}x \tag{5.10}$$

Since there are no electrical or pressure gradients in the system and no concentration gradients other than for the isotope, i, the general equations [Eqns. (5.8)] simplify to Eqn. (5.4), which is simply Fick's law. This we may now apply to three separate

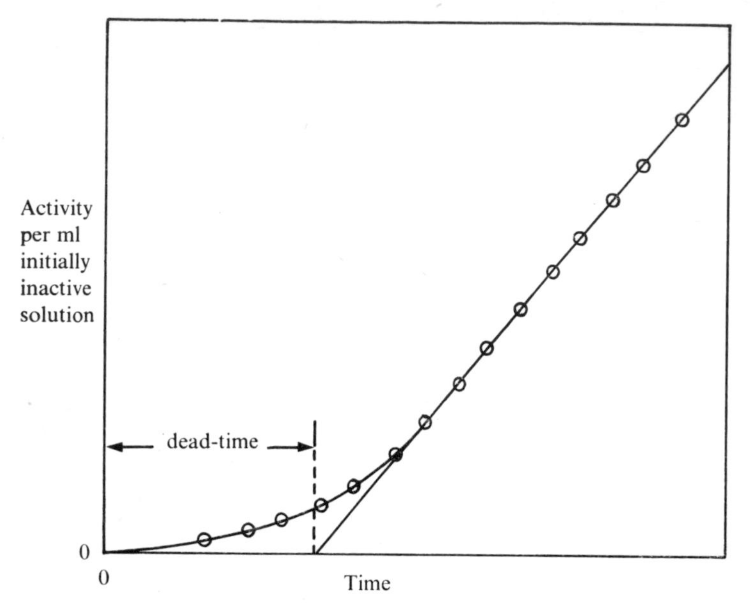

FIG. 5.3a. A plot of the accumulation of radioactivity on the initially inactive solution during an isotope diffusion experiment, illustrating the dead-time (or time lag) before a steady state of isotopic flow is achieved

diffusion steps under the steady state limitation that the flow J_i is a constant across films and membrane.

$$J_i = \frac{D\,(c_i - c_i')}{\delta} = \frac{\bar{D}\,(\bar{c}_i' - \bar{c}_i'')}{d} = \frac{D(c_i'' - 0)}{\delta} \tag{5.11}$$

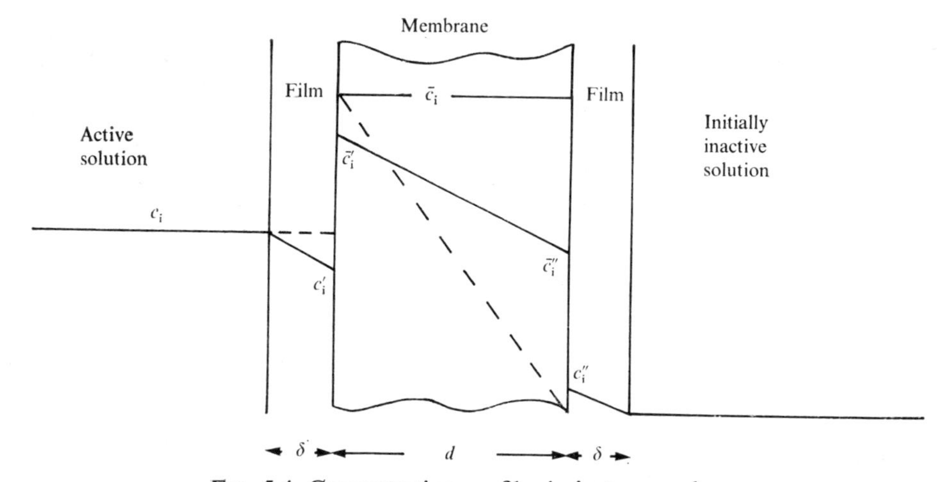

FIG. 5.4. Concentration profiles in isotope exchange

The concentrations c_i are proportional to the radioactivity per unit volume and D and $\bar{D}$ are the isotopic diffusion coefficients in solution and membrane respectively, other symbols being most clearly explained by Fig. 5.4. Since there is no interfacial resistance to diffusion at the membrane surface, equilibrium conditions may be assumed at the membrane–solution boundaries, so that

$$\frac{c_i'}{c} = \frac{\bar{c}_i'}{\bar{c}} \quad \text{and} \quad \frac{c_i''}{c} = \frac{\bar{c}_i''}{\bar{c}} \tag{5.12}$$

Elimination of c_i'; c_i'', $\bar{c}_i'$ and $\bar{c}_i''$ from Eqns. (5.11) and (5.12) yields Eqn. (5.13).

$$J_i = \frac{-\bar{D}}{1 + 2\bar{D}\bar{c}\delta/Dcd} \cdot \bar{c}/d \tag{5.13}$$

This equation has the form of Fick's simple law, in which the effective diffusion coefficient D_e given by the first terms on the right hand side of Eqn. (5.13) is directly obtained by experiment.

If $Dcd \gg 2\bar{D}\bar{c}\delta$ then $D_e = \bar{D}$ and so the overall rate of diffusion is controlled by the membrane diffusion step. Alternatively, if $Dcd \ll 2\bar{D}\bar{c}\delta$ the effective diffusion coefficient is that of the solution. Membrane control is favoured by a small $\bar{D}$, high concentration in solution, efficient stirring and using thick membranes. The ratio of diffusion coefficient in solution to membrane is invariably greater than unity, mainly as a result of the tortuous and restricted diffusion paths in the membrane matrix. The ratio $c/\bar{c}$ is also important and in the case of ions is controlled by the Donnan exclusion effect. For counter ions this ratio is small, becoming very small indeed when the external solutions are dilute and hence, particularly in this latter case, counter ion diffusion may well be film controlled. Conversely, for co-ions this ratio is usually greater than unity and increases with dilution so that film diffusion will only tend to become rate controlling under extreme conditions: more concentrated solutions and inefficient stirring. For non-electrolyte, such as solvent, the concentration ratio is of the order of unity and the diffusion is controlled by the other factors in the expression for D_e.

The experimentally determined diffusion coefficient obtained from Eqn. (5.13) is thus not, in general, a true measure of, $\bar{D}$, the diffusion coefficient in the membrane. To obtain it an estimate of δ must be made, and a variety of methods now exist in the literature.[1]

The self-diffusion coefficient depends upon the size of the solvated ion and follows the mobility series of aqueous solutions. It is, however, depending upon the membrane, about one order of ten smaller. The sequence in PSSAc membranes and beads is $Li^+ > Na^+ > K^+ > Rb^+ > Cs^+$ and as the valency of the ion increases the reduction in the diffusion coefficient becomes more pronounced: $Na^+ > Ca^{2+} > Y^{3+} > Th^{4+}$. This effect is greater for counter ions than co-ions and is less pronounced in anion exchangers. It is considered to be due to the stronger attraction of the polyvalent counter ions with the oppositely charged matrix. For co-ions this interaction is repulsive and unlikely to cause retardation.

The degree of cross-linking also controls diffusion; in general the greater the cross-linking the slower the diffusion. Meares,[2] considering this problem in detail, assumed the primary cause of the lower diffusion coefficients in the exchanger to be due to obstruction of the diffusion paths by polymer chains, causing the diffusing species to follow a much more tortuous path in the exchanger phase. From a simple statistical model he has calculated this tortuosity effect in terms of the volume fraction, V_r, of the resin in the swollen exchanger. The final relationship related the membrane diffusion coefficient, $\bar{D}$, with that of the aqueous solution, D.

$$\bar{D} = D\left(\frac{1 - V_r}{1 + V_r}\right)^2 \tag{5.14}$$

As the water content of the exchanger decreases V_r increases and accordingly $\bar{D}$ decreases also. Although purely geometric, this treatment has been useful in predicting the membrane diffusion coefficients from a knowledge of solution values. Table 5.1 gives typical results obtained by Meares[3] for Zeo-carb 315 phenol–sulphonic acid membranes.

TABLE 5.1. Self-diffusion Coefficients for Sodium and Chloride Ions in Zeo-carb 315 Phenol–Sulphonic Acid Membranes.

NaCl External Solution Concentration (moles/litre)	Self-diffusion Coefficient ($\bar{D} \times 10^6$ cm²/sec) Na⁺	Cl⁻	$\bar{D}_{obs}/\bar{D}_{calc}$ Na⁺	Cl⁻
0·01	2·33	7·50	0·51	1·07
0·02	2·97	7·46[a]	0·65	—
0·05	3·51	7·42[a]	0·77	—
0·10	3·72	7·37[a]	0·83	—
0·50	4·75	7·23[a]	1·07	—
1·00	4·70	7·08[a]	1·08	—

[a] Values calculated from Eqn. (5.14).

Typically the diffusion coefficients for counter ions increase with increasing concentration in the external solution, while for co-ions there is a steady decrease.[4,5]

In dilute solutions, at least, the fixed charge must tend to attract the counter ion, localising it near the polymer chains. The process of diffusion may then be regarded as one in which the counter ion progresses by a series of jumps between one fixed charge and the next along the polymer chains. As the concentration of the external solution increases, electrolyte invasion provides alternative negative sites and so allows less restricted diffusion paths, overriding the increasing tortuosity caused by de-swelling. For the co-ion, which is electrically repelled by the polymer fixed charges, the increase in internal electrolyte concentration will not significantly change the co-ion environment, the ratio of counter ion to co-ion remaining very large. The dominant effect will be tortuosity and so co-ion diffusion coefficients decrease as bathing solution concentrations increase. This picture appears to be qualitatively successful in strong acid cation exchangers. In anion exchangers, however, Glueckauf and Watts[6] report co-ion diffusion coefficients of sodium and lanthanum ions to increase with external concentration. This discrepancy is as yet unexplained. In general, it is more difficult to rationalise anion exchange behaviour in terms of simple concepts. There is undoubtedly a complex water interaction with quaternary ammonium fixed charges and strong possibilities of ion association with counter ions. A tentative explanation may lie in a loosening of water structure in the membrane as more external salt penetrates, allowing a freer movement of co-ions.

5-2a. Interpretation of diffusion in terms of Onsager frictional coefficients

In most systems studied the diffusion coefficient of the counter ion is smaller than that of the co-ion. An elegant confirmation of the qualitative arguments comes from the work of Meares[3] and co-workers. In an exhaustive analysis of a series of transport processes in Zeo-carb 315 phenol–sulphonic acid membrane in the sodium form, Meares has calculated the Onsager frictional coefficients for the interaction of each component with all others in the membrane. The R_{ij} coefficient represents a measure of the frictional interaction of one mole of species i with one mole of species j in its environment. The values calculated for this membrane in the sodium form bathed with 0·05M sodium chloride are given in Table 5.2.

TABLE 5.2. R_{ij} for Zeo-carb 315 in the Sodium Form at 0·5M Sodium Chloride Concentration.[a]

$-R_{13}$	$-R_{14}$	$-R_{23}$	$-R_{24}$
1·21	46·8	0·82	0·439

[a] The units of R_{ij} are (10^{-10} × joule cm sec/mole^2).

The interaction of water, 3, with sodium ion, 1, and with chloride ion, 2, are approximately equal as shown by the magnitudes of R_{13} and R_{23} respectively. On the other hand, the interaction of the matrix 4 with counter ion (R_{14}) is two orders of ten greater than R_{24}, the frictional coefficient of co-ion and matrix. The coefficients may be positive or negative. A negative sign indicates that the interaction is one in which the species under force tends to drag the other with it to some extent. The self-diffusion coefficient D_{ii} can be represented thus:[3,7]

$$D_{11} = RT/-c_3R_{13} - c_4R_{14}$$
$$D_{22} = RT/-c_3R_{23} - c_4R_{24}$$

(In this treatment the interaction of counter ion and co-ion R_{12} is taken as zero.) Where c_i is the concentration of i in the membrane expressed as moles of i per ml of resin volume. The larger diffusion coefficient for the co-ion is therefore entirely due to its smaller interaction with the matrix.

5-3. MEMBRANE POTENTIAL

When an ion exchanger is placed in an electrolyte solution an electrical potential is developed at the solution–exchanger interface known as the Donnan potential (see section 3.2). This potential varies according to the concentration and composition of the bathing solution and is defined as the difference in electrical potential between the exchanger phase and the solution. If, therefore, an ion exchange membrane separates two solutions of different concentration there is a net potential difference across the membrane as a whole. This is the membrane potential E_m.

The simplest case occurs when a membrane separates two solutions of the same salt, AX, which differ only in concentration. Assuming further that the exchanger, now in the A-form, is ideally semi-permeable we may simply derive an expression for the membrane potential.

The condition for chemical equilibrium across the system involves only the A counter ion in this case, since it is the only ion common to each phase.

$$\text{At interface I} \quad \tilde{\mu}_{A}' = \tilde{\mu}_{\bar{A}}$$
$$\text{II} \quad \tilde{\mu}_{\bar{A}} = \tilde{\mu}_{A}''$$

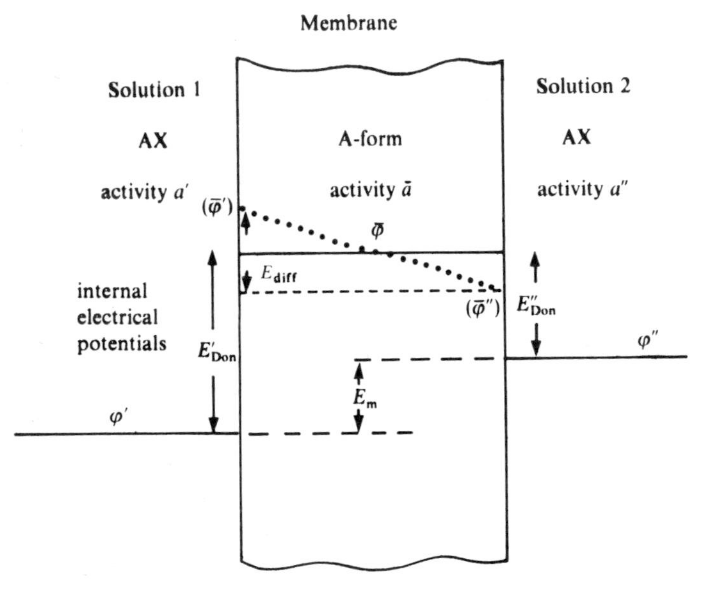

FIG. 5.5. Electrical potentials in solution and membrane developed when a membrane is bathed by solutions of AX of different concentration. For ideal membrane behaviour the internal electrical potential, φ, of the membrane phase is constant, In the more general case of non-ideal behaviour φ is no longer a constant across the membrane and a diffusion potential exists (dotted curve).

so that $\tilde{\mu}_{A}' = \tilde{\mu}_{A}''$ – the solution electrochemical potentials are equal. Expanding in terms of Eqn. (5.2a) one obtains Eqn. (5.15).

$$\mu_{\bar{A}}^{0} + RT \ln a_{A}' + z_{A}F\varphi' = \mu_{\bar{A}}^{0} + RT \ln \bar{a}_{A} + z_{A}F\bar{\varphi}$$

so that

$$E_{Don} = \bar{\varphi} - \varphi' = (\mu_{A}^{0} - \mu_{\bar{A}}^{0})/z_{A}F + \frac{RT}{z_{A}F} \ln a_{A}'/\bar{a}_{A} \tag{5.15}$$

The membrane potential is defined as the difference between the electrical potentials of the bathing solutions and so

$$E_m = \varphi'' - \varphi' = (\bar{\varphi} - \varphi') - \bar{\varphi} - \varphi'') = (E_{Don}') - E_{Don}'')$$

$$= \frac{RT}{z_A F} \ln a_A''/a_A'' \qquad \text{(see Fig. 5.5)} \tag{5.16}$$

The membrane in this ideal case behaves as an electrode reversible to A ions. If one of the solutions is kept constant in composition then

$$E_m = \text{constant} + \frac{RT}{z_A F} \ln a_A'' \tag{5.16a}$$

This is quite identical to the response of the glass electrode, which has a potential dependent upon the pH ($-\log a_{H^+}$). The analogy is not complete, since the glass electrode has a response due to hydrogen ions alone even if other cations are present in solution. No such selectivity is possible for normal ion exchange membranes and bi-ionic potentials are obtained. To obtain these ideally simple expressions for membrane potentials [Eqns. (5.16a) and (5.16b)] it is essential that $\bar{a}_A$ and $\bar{\varphi}$ are uniform across the membrane. In practice this can only be achieved if the electrolyte uptake in the membrane is small and hence electrolyte solutions must be dilute. If more concentrated solutions are used, electrolyte will penetrate the membrane to different degrees on opposite sides, causing gradients of both activity and internal electrical potentials within the membrane.

5-3a. Calculation of the membrane potential in the general condition where salt uptake is important

Once again the membrane potential is defined as the difference between the electrical potentials of the bathing solutions, and so

$$E_m = (\varphi'' - \varphi') = (\bar{\varphi}' - \varphi') + (\bar{\varphi}'' - \bar{\varphi}') + (\varphi' - \bar{\varphi}'')$$

$$E_m = E_{Don}' + E_{Diff} - E_{Don}'' \qquad \text{(see Fig. 5.5)} \tag{5.17}$$

E_{Diff}, the diffusion potential, is the difference in internal electrical potentials between the two internal faces of the membrane, and is rather analogous to the common liquid junction potential developed when two electrolyte solutions are in contact. In both cases a concentration of salt implies a gradient of chemical potential of salt and so a thermodynamic force causing diffusion. For a symmetrical electrolyte the forces on cation and anion are equal but their mobilities are in general quite different. Under equal force one or other ion will have a higher velocity. Since the ions carry charge, this partial separation of cations and anions will set up an electrical space charge across the diffusion layer and hence an electrical potential difference. The direction and magnitude of this diffusion potential E_{Diff} is such that it slows the faster ion and speeds the slower so that the ions migrate at the same rate. The result is that neutral salt diffuses, no electric current flows and electroneutrality is maintained in each solution.

The diffusion potential may be calculated from the Nernst–Planck equation or from the thermodynamics of irreversible processes, each giving the same result [Eqn. (5.18)].[8,9]

$$E_{\text{Diff}} = -\frac{RT}{z_A F}\left[\ln\left(\frac{\bar{a}_A''}{\bar{a}_A'}\right) - (z_X - z_A)\int_{\text{I}}^{\text{II}} t_X \, d(\ln a_\pm)\right] \tag{5.18}$$

$\bar{a}_A$ is the activity of A in the exchanger phase, $a_\pm$ is the mean activity of the electrolyte solution, and t_X is the co-ion transport number. It is to be noted that (once again) the ionic values Z_A and Z_X include signs and are positive for cations and negative for anions. Substituting Eqns. (5.15) and (5.18) into Eqn. (5.17) gives Eqn. (5.19).

$$E_m = -\frac{RT}{z_A F}\left[\ln\left(\frac{a_A''}{a_A'}\right) - (z_X - z_A)\int_{\text{I}}^{\text{II}} t_X \, d(\ln a_\pm)\right] \tag{5.19}$$

For an ideal semi-permeable membrane no co-ions are present in the matrix and in this case t_X is zero, reducing Eqn. (5.19) to the simple Nernst expression, given by Eqn. (5.16).

The integral or average transport number in the membrane may be obtained from measurements of the e.m.f. of a cell of the type:

$$\text{Ag}|\text{AgCl}|\text{NaCl}_{(aq)}' \;\vdots\; \begin{matrix}\text{cationic}\\ \text{membrane}\\ \text{Na-form}\end{matrix} \;\vdots\; \text{NaCl}_{(aq)}''|\text{AgCl}|\text{Ag}$$

The e.m.f. of this cell is given by the algebraic sum of the electrode potentials of the silver–silver chloride electrodes and the membrane potential, E_m.

$$E_{\text{cell}} = \Delta E_{\text{electrodes}} + E_m$$

$$= -\frac{RT}{F}\ln\left(\frac{a_{\text{Cl}''}}{a_{\text{Cl}'}}\right) - \frac{RT}{F}\ln\left(\frac{a_{\text{Na}''}}{a_{\text{Na}'}}\right) + \frac{RT}{F}2t_{\text{Cl}}^*\ln\left(\frac{a''_{\pm\text{NaCl}}}{a'_{\pm\text{NaCl}}}\right)$$

[t_{Cl}^* is the integral transport number of the chloride co-ion, the integration of Eqn. (5.19) being conducted as if t_{Cl} were a constant.]

Since $t_{\text{Na}}^* + t_{\text{Cl}}^* = 1$ (section 5.5a), and $a_\pm{}^2{}_{\text{NaCl}} = a_{\text{Na}} \cdot a_{\text{Cl}}$ (by definition) Eqn. (5.20) holds.

$$E_{\text{cell}} = -2t_{\text{Na}}^*\frac{RT}{F}\ln\frac{a_{\pm\text{NaCl}''}}{a_{\pm\text{NaCl}'}} \tag{5.20}$$

(It is easily proved that if an ideal anion exchanger separates two such solutions the cell e.m.f. is exactly zero: the potentials of the electrodes are exactly balanced by the membrane potential.)

5-4. BI-IONIC AND MULTI-IONIC POTENTIALS

Unlike the glass electrode, the ion exchange membrane electrode is not selective. When placed between solutions with more than one counter ion the resultant membrane potential is a function all the counter ions present. This has largely prevented more general use of ion exchange membrane electrodes in analysis and other applications. As before, the membrane potential is the algebraic sum of the Donnan potentials at the surfaces and the interdiffusion potential. The latter is best explained by a simple example. When two solutions AX and BX oppose one another across a membrane, an ion exchange of A and B ions occurs. Since each solution must remain electrically neutral, the number of equivalents (of charge) of A leaving solution AX must equal those of B arriving. Since these ions are, in principle, quite different as regards mobility, this balance is achieved spontaneously by the creation of an interdiffusion potential (see section 4.1), the faster ion setting up a space change of its own sign on the membrane side facing the solution with the slower counter ion. Since this interdiffusion is rapid in comparison to the salt diffusion of mono-ionic potentials, there is the further possibility that the kinetics of the exchange will be affected by film diffusion control, with the resultant effect upon the Donnan potentials. Further detail is beyond the scope of this book and the interested reader should consult Professor Helfferich's text on ion exchange.[8]

5-5. ELECTRICAL PROPERTIES OF MEMBRANES

5-5a. Transport of electric current

An ion exchanger conducts electricity by ionic flow in the same manner as an electrolyte solution. Since only counter ions and co-ions are mobile, they are the sole current carriers. Furthermore, since co-ion constitutes a very minor component, the counter ion carries the larger fraction of the current. For an ideal semi-permeable membrane, no co-ion is present by definition and thus the total ionic current is due to counter ions. Since the transport number is the fraction of the total current carried by an ion, it follows that the counter ion in such a membrane would have a transport number of unity regardless of the type, valency, or mobility of that ion. The natural frame of reference for flows is the stationary membrane (a membrane-fixed frame of reference). Electric currents are measured as current densities in amp/cm^2 of membrane area and I, the current density, is related to the molar flows of ions by Eqn. (5.21), the

$$I = (z_A J_A + z_X J_X) \cdot F \tag{5.21}$$

summation being taken, in the general case, over all mobile ions present in the membrane. z and J are the ionic valency, including sign, and the flux of ion in ions/cm^2 sec, respectively. F is the Faraday of electricity measured in coulombs of charge/ion equivalent. Dimensionally the right-hand side of Eqn. (5.21) has the form:

$$zJF = \frac{\text{valency . g . ion . coulomb}}{\text{cm}^2\text{ . sec . equivalent}} = \frac{\text{coulomb}}{\text{cm}^2\text{ sec}} = \frac{\text{amp}}{\text{cm}^2}$$

The equation also shows that positive and negative ions, although migrating in opposite directions in an electric field, contribute positively to the current, since $z_i J_i$ is always positive.

5-5b. Transport and transference numbers

The transport number, as defined above, may be written as in Eqn (5.22).

$$t_A = \frac{z_A J_A F}{I} = \frac{z_A J_A}{z_A J_A + z_X J_X} \tag{5.22}$$

$$t_i = \frac{z_i J_i}{\sum_{i=1}^{i=n} z_i J_i} \tag{5.23}$$

Equation (5.23) shows this in a general form, for summation over all mobile ions.

$$\left(\text{Note} \sum_{i=1}^{i=n} t_i = 1\right)$$

The transport number will therefore be a function of the ionic form of the exchanger and the proportion of invading co-ion. It is now more obvious why the transport number obtained from e.m.f. measurements is the average or integral value, since the transport number will not be constant across a membrane separating solutions of different concentration, for these will invade the membrane pores to different degrees.

To accommodate the situation (as in electro-osmosis) where solvent or other neutral species, O, is transported with ionic current, the transference number is defined as the number of moles of species transferred per faraday of electricity per unit area of membrane in the direction of positive ion current.

$$T_O = \frac{J_O F}{(z_A J_A + z_X J_X)F} \tag{5.24}$$

For an ion T is simply the transport number of the ion divided by z_i.

Unlike the transport number this has a real non-zero value for an uncharged species. It has a positive sign for component transferred with the positive ion and negative one when it is transferred with the negative ion.

5-5c. Electrical conductivity

The specific conductivity of an ion exchange membrane is the reciprocal of the electrical resistance of a one centimetre cube of the material measured across opposite faces. The electrical conductivity of exchangers is high because the internal concentration of ions in the aqueous pores is high, in some cases approaching seven molal. Conductivity increases with increasing capacity of the matrix and is decreased by increasing cross-linking and reduced water content and so by increased tortuosity (discussion, section 5.3). Although the transport number is primarily a function of the degree of semi-permeability of the matrix the conductivity is largely a function of ionic mobility. In strong acid exchangers the conductivity is greatest for the hydrogen form and follows the sequence observed in aqueous solution for other ions, unless specific interaction between counter ion and fixed charge to form ion-pairs is important.

Conductivity has been measured in a variety of different ways using both direct and alternating current. A simple example[10] involves a specially constructed conductivity cell consisting of two separate halves, each containing one large platinum electrode (Fig. 5.6).

The membrane, in the form of a disc, is clamped between the half-cells, the whole apparatus filled with the equilibrium electrolyte. The total resistance of the cell (R_t) is measured. The experiment is then repeated with no membrane present, in order to find R_s. Since the solution and membrane are in series, $R_t = R_s + R_m$ where R_m is the resistance of the membrane area A exposed. The specific conductivity k is then

$$k = \frac{1}{R_t - R_s} \cdot \frac{d}{A}$$

d is the membrane thickness.

Since the membrane resistance is obtained by difference, the area of the membrane exposed is reduced to a minimum, so that the highly conducting membrane becomes a significant proportion of the total resistance of the cell. For an accuracy of one per cent or better the cell must have a thermostat (the temperature coefficient of conductivity is of the order of two per cent per degree centigrade). If the membrane is relatively thick, there is also a significant contribution to the total conductivity due to conductivity paths around the edges of the exposed membrane areas, leading to

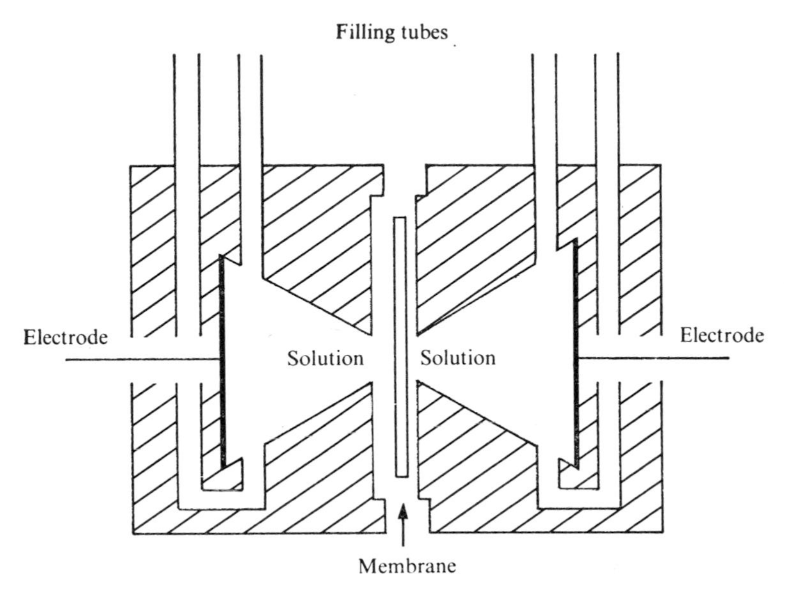

FIG. 5.6. Conductivity cell

lower resistances. Satisfactory corrections for this edge effect have been calculated by Barrer.[11]

5-5d. Measurement of transport numbers

Two methods are currently used. The first is by measuring the e.m.f. of a concentration cell and calculating the transport number from the observed membrane potential (see section 5.4). This method has the advantage of speed and simplicity, but the transport number obtained is an integral or average value across the membrane. If, however, the bathing solutions do not differ substantially in concentration the method is satisfactory. The second is the Hittorf method, in which direct current is passed and the amount of ion transported per coulomb of electricity obtained by direct analysis of the solutions. This too suffers from the disadvantage that a large number of ions must be transported for a classical solution analysis, and the resultant changes of concentration set up membrane potentials which oppose the externally applied potential differences at the electrodes. To a large degree this can be overcome by using radioactive tracer methods. Essentially, the method is identical to that used in self-diffusion (section 5.3) except that the electric potential is applied from reversible electrodes placed on either side of the membrane under test. The steady state fluxes of isotope in each direction are measured and the net ion flux obtained by difference. The advantage of this method is that bulk concentration differences may be reduced to one per cent or less, so that the membrane is effectively in contact with solutions of constant composition throughout the measurement. More serious, however, are the effects of current density upon the apparent transport number measured. Particularly in dilute solutions, increasing the current density of the measurement gives decreasing values of transport number, due to polarisation at the membrane surface.

If the membrane in the A-form is bathed by two identical solutions AX, then the current density I is made up of cation and anion components in solution, and almost solely by counter ion in the membrane.

$$I_s = (z_A J_A + z_X J_X)F \quad \text{(in solution)} \tag{5.25}$$

$$I_m = z_A \bar{J}_A F \quad \text{(in membrane)} \tag{5.26}$$

If the cross-sectional areas of the solution and the membrane are equal, $I_s = I_m$, and hence $\bar{J}_A > J_A$. More A ions leave the surface of the exchanger than can be

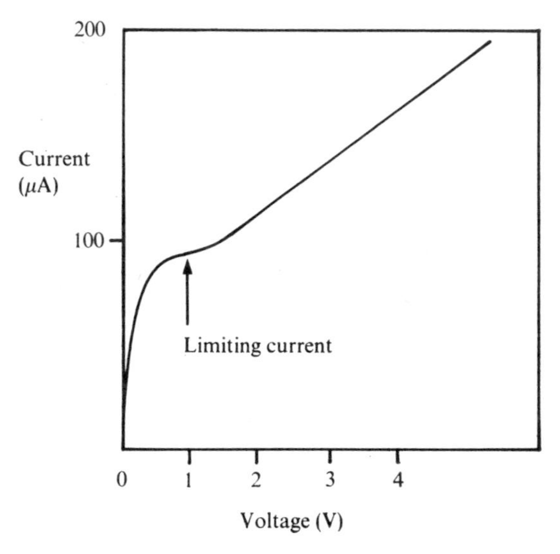

FIG. 5.7. Current–voltage characteristics of Amberplex C-10 cation exchanger between 0·01 molar sodium chloride solutions. Membrane area 0·126 cm². (Taken from reference 12, with permission)

supplied by transport in the solution. The result is that concentration gradients occur at the membrane surface. These cannot be entirely removed by stirring the solutions since it is impossible to efficiently stir right up to a solid–solution interface. If the current density is further increased the concentration of salt on the in-going side will approach zero. The resistance of the solution layer increases (and so the voltage drop) until hydrogen ions (or hydroxyl ions, in the case of an anion exchanger) are formed to provide current carriers in the membrane. The critical current density is the almost constant value obtained before water dissociation becomes important (Fig. 5.7).

For a dilute solution of a 1 : 1 electrolyte (≈0·01M) the critical current density may be as low as 20 mA/cm². Efficient agitation is required in all applications where membranes are used in electrolysis techniques. In particular it is a limiting feature of the electrodyalitic method for demineralisation of salt solutions (section 5.8).

5-6. ELECTROKINETIC PHENOMENA

5-6a. Electro-osmosis and electro-osmotic pressure

When a current is passed through a membrane, no direct electrical force is applied to the water solvent in the pores of the exchanger since it is electrically neutral. Water is nevertheless affected indirectly by ion–water dipole attraction. Since the

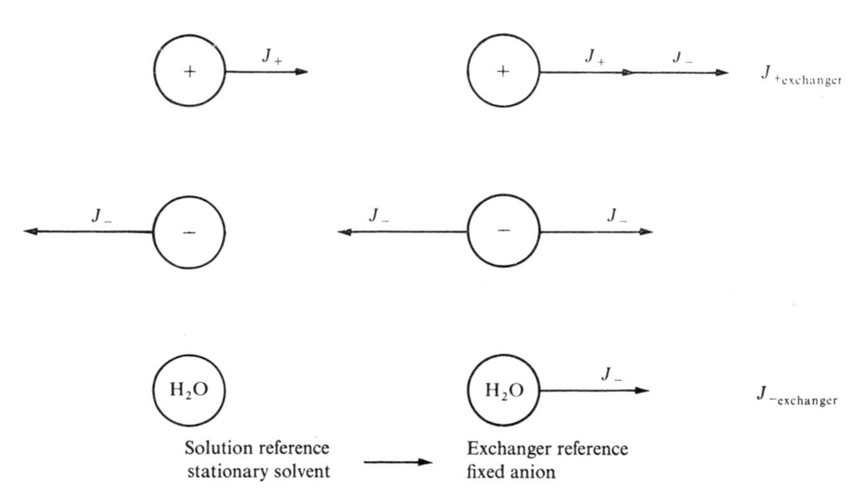

FIG. 5.8. A comparison of the flows of ions and water on a solvent-fixed reference frame (as in solution) and on an ion-fixed frame of reference (as is usual in membrane studies)

flow of counter ion is much greater than co-ion (due to their relative abundances in the matrix) there is a net flow of water with the counter ion (Fig. 5.2).

The situation is equally well explained by considering an ionic solution in an electrical field. The current is carried by both cations and anions and the usual frame of reference is the stationary solvent (Fig. 5.8), lengths and directions of the size and directions of flows being in the normal manner for vectors. To convert to fixed anion as reference, a 'back' flow is applied to all species in (a) equal in size and opposite in direction to the anion. The anion is therefore stationary, the cation flow enhanced and the water transported in the same direction as the counter ion. Electro-osmosis is therefore not a new phenomenon, but one which is described and observed because the natural frame of reference for motion in membrane experiments is the fixed membrane itself.

If the solvent flow of electro-osmosis is channelled into a vertical tube, there is a resultant increase in pressure until the solvent flow due to electro-osmosis equals the reverse flow of solvent due to the pressure head. The static pressure so developed is the electro-osmotic pressure.

In practice this can only be achieved in rather open-structured exchangers where the water permeability is high. It is typically observed in leaky membranes, porous plugs and capillaries.

5-6b. Streaming current and streaming potential

When a membrane separates two identical solutions there is no membrane potential. Applying excess pressure to the solution on one side of the membrane forces the pore liquid through the membrane (Fig. 5.9). Displacement of the counter ion, caused by movement of pore liquid, sets up an electrical potential across the membrane; the streaming potential. This potential acts upon the counter ions of the pore liquid,

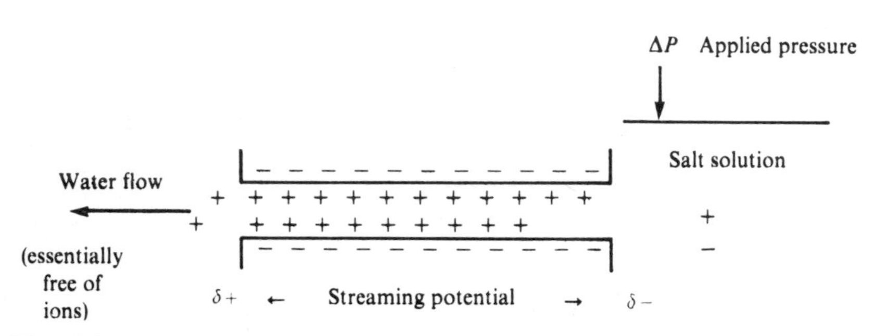

FIG. 5.9. A diagrammatic representation of the source of streaming potential

maintaining the overall electroneutrality of each phase and preventing counter ion flow into the low pressure solution. At the same time it aids co-ion flow. The net effect is that the emergent solution is considerably more dilute than the bulk with a salt content determined by the co-ion flow.

Ion exchange membranes therefore behave under pressure as ionic sieves, allowing water to pass but not salt. Known as hyperfiltration (or less usefully as reverse osmosis) this effect is used as a method of de-ionising water.

If the streaming potential is short-circuited by placing two electrodes in contact with opposite faces of the membrane, the current that flows in the external circuit is known as the streaming current.

5-7. QUANTITATIVE RELATIONSHIPS BETWEEN ELECTROKINETIC PHENOMENA

Irreversible thermodynamics may be applied in this case, and phenomenological equations may be written using a relatively free choice of flows and forces within the limits laid down by theory.[13] A particularly useful transformation defines two flows: I, the current density and J_v the volume flow (ml/cm^2) through the membrane and two forces, the negative gradients of electrical potential and pressure across the membrane: $-\text{grad}\,\varphi$ and $-\text{grad}\,P$ respectively. The phenomenological equations now become Eqns. (5.27) and (5.28).

$$I = L_{11}(-\text{grad}\,\varphi) + L_{12}(-\text{grad}\,P) \tag{5.27}$$

$$J_v = L_{21}(-\text{grad}\,\varphi) + L_{22}(-\text{grad}\,P) \tag{5.28}$$

Once again the Onsager reciprocal relations hold: $L_{12} = L_{21}$. L_{11} is the specific conductivity, since if grad P is zero, Eqn. (5.27) becomes Eqn. (5.29), which is another way of expressing Ohm's law.

$$I = L_{11}(-\text{grad}\ \varphi) \tag{5.29}$$

L_{22} is the flow permeability of the membrane, since if grad φ is zero Eqn. (5.28) becomes Eqn. (5.30).

$$J_v = L_{22}(-\text{grad}\ P) \tag{5.30}$$

The direct effect of an electrical potential gradient is to cause current flow – indirectly it will cause volume flow. Explained qualitatively above in section 5.6, such effects are given quantitative meaning by Eqns. (5.27) and (5.28).

$$J_v = L_{21}(-\text{grad}\ \varphi) \quad \text{if} \quad \text{grad}\ P = 0 \tag{5.31}$$

and similarly

$$I = L_{12}(-\text{grad}\ P) \quad \text{if} \quad \text{grad}\ \varphi = 0 \tag{5.32}$$

A flow will not disappear when the direct force vanishes unless there is no coupling of flows. Under these conditions $L_{12} = L_{21} = 0$, and all electrokinetic phenomena would disappear.

5-7a. Electro-osmosis

The conditions for electro-osmosis are that grad $P = 0$ and I, grad φ and $J_v \neq 0$.

From Eqns. (5.27) and (5.28) the electro-osmotic flow is given by Eqn. (5.33).

$$(J_v/I)_{\text{grad}\ P=0} = L_{21}/L_{11} \tag{5.33}$$

J_v/I is a measure of the volume flow per coulomb of electricity flowing and is defined as the electro-osmotic permeability.

5-7b. Streaming potential

Under conditions of zero current an applied pressure gradient will result in the development of an electrical potential in addition to expected volume flow.

$$I = 0 \quad \text{and} \quad \text{grad}\ P,\ J_v \text{ and grad}\ \varphi \neq 0$$

and so from Eqn. (5.27), Eqn. (5.34) it follows that

$$(\text{grad}\ \varphi/\text{grad}\ P)_{I=0} = (\Delta E/\Delta P)_{I=0} = -L_{12}/L_{11} \tag{5.34}$$

ΔE and ΔP are the electrical potential and pressure differences across the membrane.

From the Onsager reciprocal relations $L_{12} = L_{21}$ Eqn. (5.35) becomes obvious.

$$(J_v/I)_{\text{grad}\ (P=0)} = -(\Delta E/\Delta P)_{(I=0)} \tag{5.35}$$

The electro-osmotic permeability equals the negative of streaming potential per unit pressure difference.

5-7c. Electro-osmotic pressure and streaming current

The conditions for the development of electro-osmotic pressure are that under an electrical potential gradient and with zero volume flow there will develop a static pressure gradient.

From Eqn. (5.28) with $J_v = 0$ and grad φ, I and grad $P \neq 0$.

$$(\text{grad } P/\text{grad } \varphi)_{J_v=0} = (\Delta P/\Delta E)_{J_v=0} = -L_{21}/L_{22} \quad (5.36)$$

The conditions for streaming current are grad $\varphi = 0$ and J_v, grad P and $I \neq 0$. Substituting in Eqns. (5.27) and (5.28) an expression for the streaming current in terms of L coefficients is obtained [Eqn. (5.37)].

$$(I/J_v)_{\text{grad } \varphi=0} = L_{12}/L_{22} \quad (5.37)$$

Again from the Onsager reciprocal relations, Eqn. (5.38) can be obtained.

$$(I/J_v)_{\text{grad } \varphi=0} = -(\Delta P/\Delta E)_{J_v=0} \quad (5.38)$$

The streaming potential/unit volume flow equals the (negative) electro-osmotic pressure per unit applied voltage. Equations (5.35) and (5.38) describe the relationships between the electrokinetic phenomena. They are known as Saxen's relations in honour of the nineteenth-century chemist who first observed them experimentally. It is obvious that these phenomena are best described by the irreversible thermodynamic treatment, and further that they depend upon a non-zero value of the coupling coefficient L_{12}.

5-8. COMPOSITE MEMBRANE SYSTEMS

5-8a. Two or more membranes in series

Consider a simple system in which two ideal semi-permeable membranes separate three identical chloride solutions (Fig. 5.10). If electrical potential is applied to two reversible silver–silver chloride electrodes placed in the outer solutions electric current will flow.

Case 1: If the left-hand electrode is negative, the net effect of current will be to accumulate salt in the central compartment and deplete concentration in 3. Since the membranes are taken as ideally semi-permeable they behave as single ion conductors.

Case 2: With the left-hand electrode positive the net effect will be the reverse of Case 1. Salt will be removed from the central compartment and increased in compartment 3. The latter represents a simple and practical way of de-ionising water and is known as *electrodialysis*. Extending the system to a large bank of membranes which are alternately cationic and anionic, it is obvious that the effect of current will be to de-ionise and concentrate solution in alternate chambers.

As a practical and economical method for de-ionising water or alternatively for concentrating valuable effluent, it suffers from some serious disadvantages.

The first is that, since the electrical resistance of pure water is high, a large electrical potential is required to remove the last traces of ions. As the de-ionising proceeds, the concentration gradients of salt increase across the membrane, tending to return salt by back diffusion into the dilute solutions. Secondly, these concentration gradients generate membrane potentials which oppose the applied potential, further reducing the electric current in the cell.

Current densities must also be kept below their limiting value (see section 5.5c). Finally, electro-osmotic transport of water is always out of the dilute solution and into the concentrated one, further reducing efficiency owing to the electrical work done in this process.

The most efficient units would therefore maintain inter-membrane distances as small as possible to keep resistance low. They should be efficiently stirred, reducing

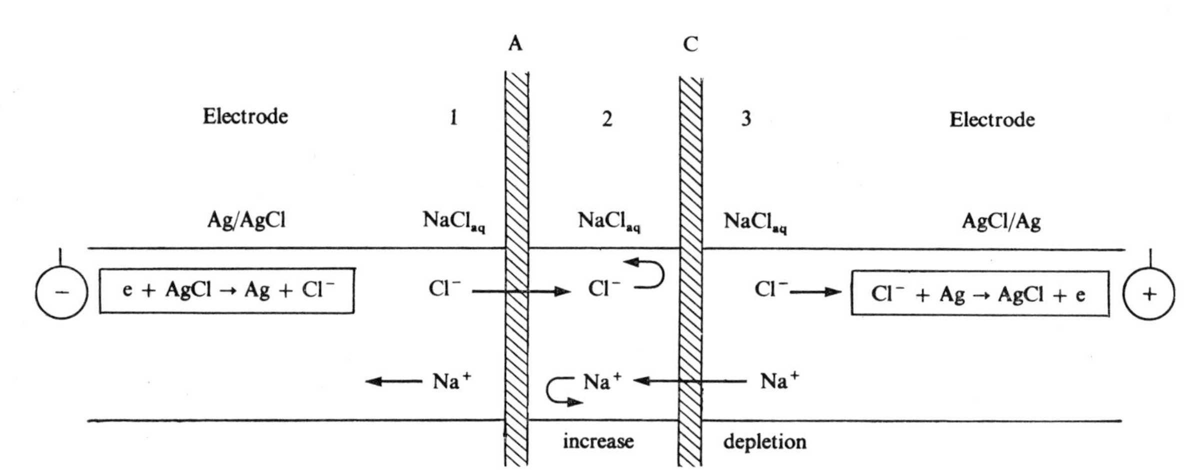

FIG. 5.10. Composite membrane systems; electrodialysis in a cell containing cation and anion exchange membranes in series

concentration gradients at the membranes and increasing the threshold of the limiting current. These limitations are sufficiently great that a high degree of plant sophistication must be incorporated before it may be used for the large scale purification of water, although its simplicity renders it a useful method in laboratory work or in conditions where expensive ions are to be recovered and cost is less important.

5-8b. Membranes in parallel: the Sollner ring

The permeability of a cation and anion exchanger in parallel (Fig. 5.11) is much greater than each membrane considered separately.

If a cation or anion exchanger separates solutions of different concentration, no electric current will flow and only salt (and water) will diffuse. If placed in parallel,

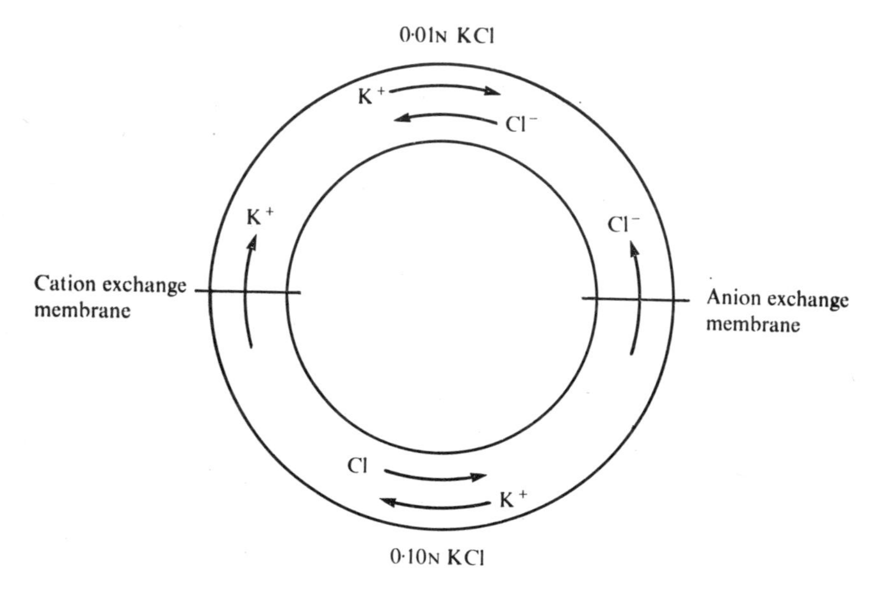

FIG. 5.11. The Sollner ring: spontaneous flow of ionic current in a system containing two membranes in parallel

electroneutrality in each solution may be maintained if the flow of positive ion through the cation and negative ion through the anion exchanger carry equal numbers of charges. The spontaneous tendency for the concentrations to equalise is allowed by this process, which amounts effectively to a spontaneous flow of electric current around the ring. This is a striking example of electric current in a system without electrodes.

5-9. ACTIVE TRANSPORT

One of the greatest challenges in biology is the detailed understanding of active transport. Active transport is the transport of an ion or molecule in a direction opposed to the normal direction for diffusional flow, i.e. against electrical or concentration gradients. In biology such systems are known as pumps; sodium, potassium and calcium pumps are well known.

Within the context of the discussion so far this understanding is impossible. The concept which has to be introduced is the chemical affinity. This is the force which drives a chemical reaction to completion and it is the coupling of chemical forces and diffusional flows which results in active transport. Consider reaction (5.39), in which x molecules of X react with y molecules of Y to form z molecules of a product Z.

$$xX + yY \rightarrow zZ \tag{5.39}$$

At equilibrium the free energies of the reactants and products are equal. No net chemical reaction is in progress and so the driving force for chemical reaction is zero.

At equilibrium

$$x\mu_X + y\mu_Y - z\mu_Z = 0 \tag{5.40}$$

Near equilibrium the reactants have a higher free energy than the products and the reaction proceeds from left to right. The driving force for the chemical reaction is the affinity A, defined as in Eqn. (5.41).

$$x\mu_X + y\mu_Y - z\mu_Z = A \tag{5.41}$$

It follows that A is zero when the system is in chemical equilibrium. The flow corresponding to A is the rate of chemical reaction, J_{ch}. There is one major difference between these and the diffusional forces and flows, and that is that they are scalars. They have no direction in space, unlike gradients of chemical potential and diffusional flows, which are vectors.

It is well-known fact that homogeneous chemical reactions in solution cannot set up concentration or thermal gradients in the reaction mixture. In terms of irreversible thermodynamics this means that there can be no coupling between the chemical reaction and diffusional or thermal flows. This is a specific example of the Curie–Prigogine principle,[15] which states that a scalar force (the affinity in this case) cannot produce a vector flow in a isotropic system. It is therefore necessary that any system which involves active transport (and so uses a chemical reaction to produce a vector flow in ions) must be asymmetric in the direction of flow. In a membrane system, the principle requires that active transport may only occur across an asymmetric membrane if the bathing solutions are identical.

A classic example of active transport of sodium ion occurs with the frog skin. If the skin is placed between two identical solutions of chloride Ringer's solution there is a selective transport of sodium ion from outside to inside. Since the solutions are identical, there can be no gradients of chemical potential of sodium ion. The membrane therefore acts as a pump for sodium ions. It contains within it a highly sophisticated pore system together with the requisite chemical (metabolic) reactions which are used as a driving force for the sodium ions. The Curie–Prigogine principle also requires that the membrane be asymmetric, its outside and inside surfaces differing in chemical and/or physical structure. The mechanisms of such biological systems remain unsolved. This is not surprising, since natural membranes are most complex and the physical nature of the membrane and the metabolic driving reaction are essentially unknown.

A common assumption in the theories of active transport is that the ion transported is at some stage attached to an activated carrier molecule which is permeable to the membrane in a way that the free ions are not. On passing the membrane this is deactivated, releasing the free ion. In this way a selective carrier will provide a selective transport of ions. In a simple physical model system no attempt can be made to parallel the specificity of the natural membranes. The object of a model system is simply to construct a pump mechanism for an ion using a known chemical reaction as a driving force, a specific carrier mechanism and membranes of known characteristics. One such system will now be considered.[16]

In outline the model pump consists of a two membrane system and contains three solutions, a, b, and c. Membrane I is a simple cation exchange membrane which is effectively impermeable to anions. The cation to be pumped, A, is in the simple aquo or uncomplexed form in solutions a and c. In b, however, it is complexed by the anion of a weak acid with which it forms an uncharged species AX. There is thus set up a concentration gradient for the A ion across membrane I, driving this ion into solution

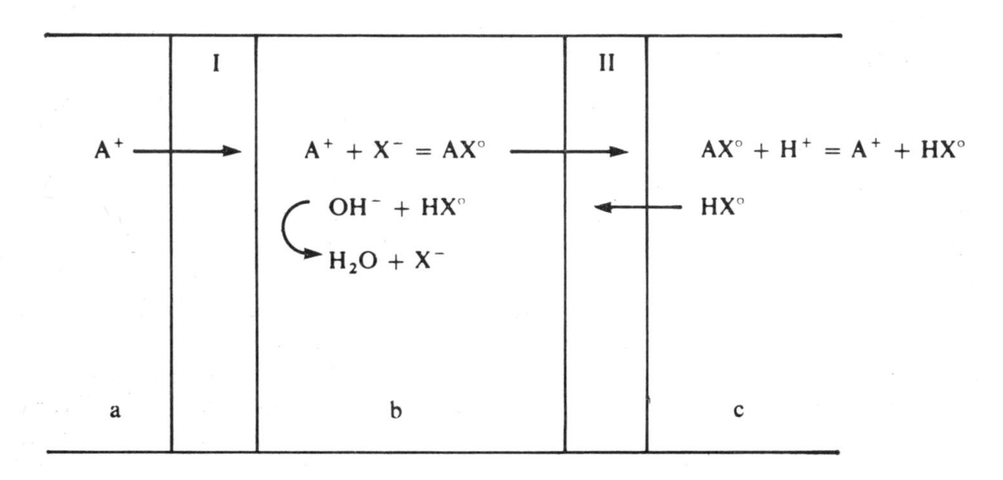

FIG. 5.12. A model system for active transport

b. When A passes into solution b it is complexed by an excess of the ligand X so that effectively no free A is present. Membrane II is designed to allow AX to pass into solution c while retaining the anionic ligand X. It is therefore a cation exchanger on the side facing solution b. On passing II by diffusion into solution c the complex is destroyed by strong acid, releasing the free ion A and producing the associated weak acid HX. To prevent back diffusion of A and free hydrogen ion membrane II must be an anion exchanger on the side facing solution c. It must therefore be a cation–anion exchanger sandwich. Since each component of the sandwich is separately impermeable to either anion or cation, the composite membrane will be impermeable to both, while still allowing the passage of neutral species which are not affected by the Donnan potentials. The membranes therefore act as one-way valves, allowing only those ions involved in 'active' transport to pass, and thus retaining in b the complexing ligand and in c the transported A and hydrogen ions.

As the system operates, a concentration of associated HX builds up in solution c and back diffusion of this acid into b will become important. (Again it is assumed that membrane II is impermeable to ions but not to uncharged molecules.) If we consider a system in which the pH's of b and c are maintained, a steady-state will be reached such that the concentrations of HX in c and anion X in b will remain constant. In such a system the net chemical reaction will be the reaction of hydrogen and hydroxyl ions to form water. The situation is summarised in Fig. 5.12. This mechanism has been tested using the cupric ion in place of A, acetate and acetic acid for X and HX respectively and $CuAc_2$ as the neutral complex carrier.

REFERENCES

1. E. M. Scattergood and E. N. Lightfoot, *Trans. Faraday Soc.* **64,** 1135 (1968).
2. J. S. Mackie and P. Meares, *Proc. Roy. Soc.* (*London*) *Ser. A.* **232,** 485, 498 and 510 (1955).
3. D. Mackay and P. Meares, *Trans. Faraday Soc.* **55,** 1221 (1959).
4. N. Ishibashi, T. Seiyaina and W. Sakai, *J. Electrochem. Soc.* (*Japan*) **27,** E-193 (1959).
5. R. Schogl, *Z. Electrochem.* **57,** 195 (1953).
6. E. Glueckauf and R. E. Watts, *Proc. Roy. Soc.* (*London*) *Ser. A.* **268,** 339 (1962).
7. R. W. Laity, *J. Phys. Chem.* **63,** 80 (1959).
8. F. Helfferich, *Ion Exchange*, McGraw-Hill, New York, 1962.
9. D. G. Miller, *J. Phys. Chem.* **70,** 2645 (1966).
10. *Test Manual* (*Tentative*) *for Permselective Membranes*, R. and D. Report No. 77, U.S. Office of Saline Water.
11. R. M. Barrer, J. A. Barrie and M. G. Rogers, *Trans. Faraday Soc.* **58,** 2473 (1962).
12. A. M. Peers, *Discussions Faraday Soc.* **21,** 124 (1956).
13. A. Katchalsky and P. Curran, *Non-Equilibrium Thermodynamics in Biophysics*, Harvard University Press, Cambridge, Mass., 1965.
14. K. Sollner, S. Dray, E. Grim and R. Neihof in *Electrochemistry in Biology and Medicine* (T. Shedlovsky, Ed.), John Wiley and Sons, New York, 1955.
15. D. D. Fitts, *Non-Equilibrium Thermodynamics*, McGraw-Hill, New York, 1962.
16. R. Paterson, *Nature* **217,** 545 (1968).

FURTHER READING

F. Helfferich, *Ion Exchange*, McGraw-Hill, New York, 1962, Chapter 8.

A. Katchalsky and P. Curran, *Non-Equilibrium Thermodynamics in Biophysics*, Harvard University Press, Cambridge, Mass., 1965.

Ion Transport Across Membranes (Hans T. Clarke, Ed.), Academic Press, New York, 1954.

Membrane Transport and Metabolism (O. Kedem, Ed.), (*Proceedings of a Symposium held in Prague*, 1960; A. Klienzeller and A. Kotyk, Eds.), Academic Press, New York, 1961.

N. Lakshminarayanaiah, 'Transport Phenomena in Artificial Membranes', *Chem. Rev.* **65,** 491 (1965). (An excellent review with 609 references.)

S. R. Caplan and D. C. Mikulecky, 'Transport Processes in Membranes', in *Ion Exchange* (J. A. Marinsky, Ed.), Marcel Dekker, New York, 1966.

6 Column techniques and chromatography

The simplest and most common application of ion exchange is the conversion of a salt solution of BY into AY by ion exchange of B ion with the A-form of an exchanger. The process may be represented by an equilibrium:

$$BY_{aq.} + \overline{A{-}R} \rightleftharpoons \overline{B{-}R} + AY_{aq.}$$

For example:

$$NaCl_{aq.} + \overline{H{-}R} \rightleftharpoons \overline{Na{-}R} + HCl_{aq.}$$

$$NaCl_{aq.} + \overline{OH{-}R} \rightleftharpoons \overline{Cl{-}R} + NaOH_{aq.}$$

In batch experiments, the degree of conversion will depend upon the selectivity of the exchanger for the ions ($K_A{}^B$) and the relative amount of the exchangeable ions in the solution and exchanger batch. A simple example illustrates this more clearly.

1 ml of wet resin in the hydrogen form (capacity 2 mequiv/ml) is equilibrated with 10 ml of sodium chloride solution, 0·1N. The total exchangeable sodium in solution is 10 × 0·1/1000 equivalents or more simply, 1 milli-equivalent (mequiv).

If x mequiv of sodium are exchanged with the resin at equilibrium, then Eqn. (6.1) holds.

$$K_H{}^{Na} = 2 = \frac{[\bar{Na}]\cdot[H]}{[\bar{H}]\cdot[Na]} = \frac{\left(\dfrac{x}{v_r}\right)\left(\dfrac{x}{v_s}\right)}{\left(\dfrac{2-x}{v_r}\right)\left(\dfrac{1-x}{v_s}\right)} = \frac{x^2}{(2-x)(1-x)} \tag{6.1}$$

Square brackets represent molar concentrations, while v_r and v_s are the sample volume of resin and solution respectively.

Solving for x in Eqn. (6.1) gives $x = 0{\cdot}763$, showing that 76% of the sodium ions are exchanged with hydrogen ions in this equilibration. Although the selectivity is in favour of sodium in this exchange, it is obvious that several equilibrations with fresh sodium chloride solution would be required for a complete conversion of the exchanger to the sodium form. This method of conversion would obviously be troublesome and tedious, and in practice column techniques are used.

An ion exchange column contains exchanger in a close-packed column in which the interstices are entirely filled with solution or solvent. Solution for exchange enters the top of the column as feed and is withdrawn as effluent. If a sample of BY is placed on the top of a column in the A-form and the solution allowed to percolate slowly through the beads, a series of equilibrations occur between B and A on successive layers of the column. If the solution BY is fed continuously on the column, the top layers are continuously bathed with fresh BY and are soon converted entirely into the B-form. The level at which exchange occurs in the column is displaced downwards and eventually, when the column is almost entirely converted into the B-form, B ions will appear in the effluent.

The volume of effluent collected up to this point is the breakthrough volume of the column and the amount of A ions displaced is the breakthrough capacity. The degree of utilisation of the column is the ratio of the breakthrough capacity to the total capacity of the column (Fig. 6.1).

If the ion B is selected preferentially over A, the exchange of B is favourable and the column is efficiently converted into the B-form; the degree of column utilisation is high. A ions are efficiently displaced and poorly retained in the exchanger phase, even when the ion B is a minor component of the solution. The effect is such that the concentration profile in the column is sharp and self-sharpening. Such a process is a displacement.

If the feed ion is not preferred, the process is one of elution. Under these conditions, B ions compete unfavourably for sites on the exchanger. A ions are retained even when the bathing solution is predominantly BY. The result is that B ions penetrate far down the column before they are all exchanged. (The degree of column utilisation is low.) The concentration profile is diffuse and non-sharpening, becoming more and more diffuse as exchange extends further and further down the column.

6-1. SIMPLE COLUMN OPERATIONS AND PRELIMINARY CONDITIONING OF RESIN EXCHANGERS

Commercial resin exchangers are usually supplied in a variety of particle sizes. The most common for general use are 50–100 mesh (0·29–0·14 mm in diameter). These may contain iron and monomer when obtained directly from the manufacturer unless they are of analytical grade or are in some way guaranteed.

A suitable conditioning procedure consists in putting the resin through a number of cycles by a series of equilibrations with hydrochloric acid (1–2M) and sodium hydroxide with intermediate washings with water and alcohol. The alcohol is used to leach out monomer. Washing is best carried out by decanting, when fines and colloidal material are removed. Subsequent conversion into any suitable form is achieved by column washing.

Typical columns for simple laboratory use consist of a glass column some 10 to 30 cm long with internal diameter around 1 cm. The column is first filled with distilled water and the resin, in a slurry with water, is poured into it, settling to the bottom in even layers. Considerable care must be taken to prevent air bubbles from being trapped in the resin bed.

A simple application of such a column is the conversion of a salt into an equivalent amount of acid by exchange with the hydrogen form of a cation exchange resin. For a general purpose PSSA exchanger (such as Amberlite IR-120) the capacity is around 2 mequiv/ml of wet resin in the H-form. A column containing 10 ml of resin will have capacity 20 mequiv. To convert the column to the hydrogen form a solution of 2N HCl is passed for some time. (Around 100 ml containing 2 × 100 mequiv of H^+ ions should be sufficient if passed through slowly.) The column is then washed free

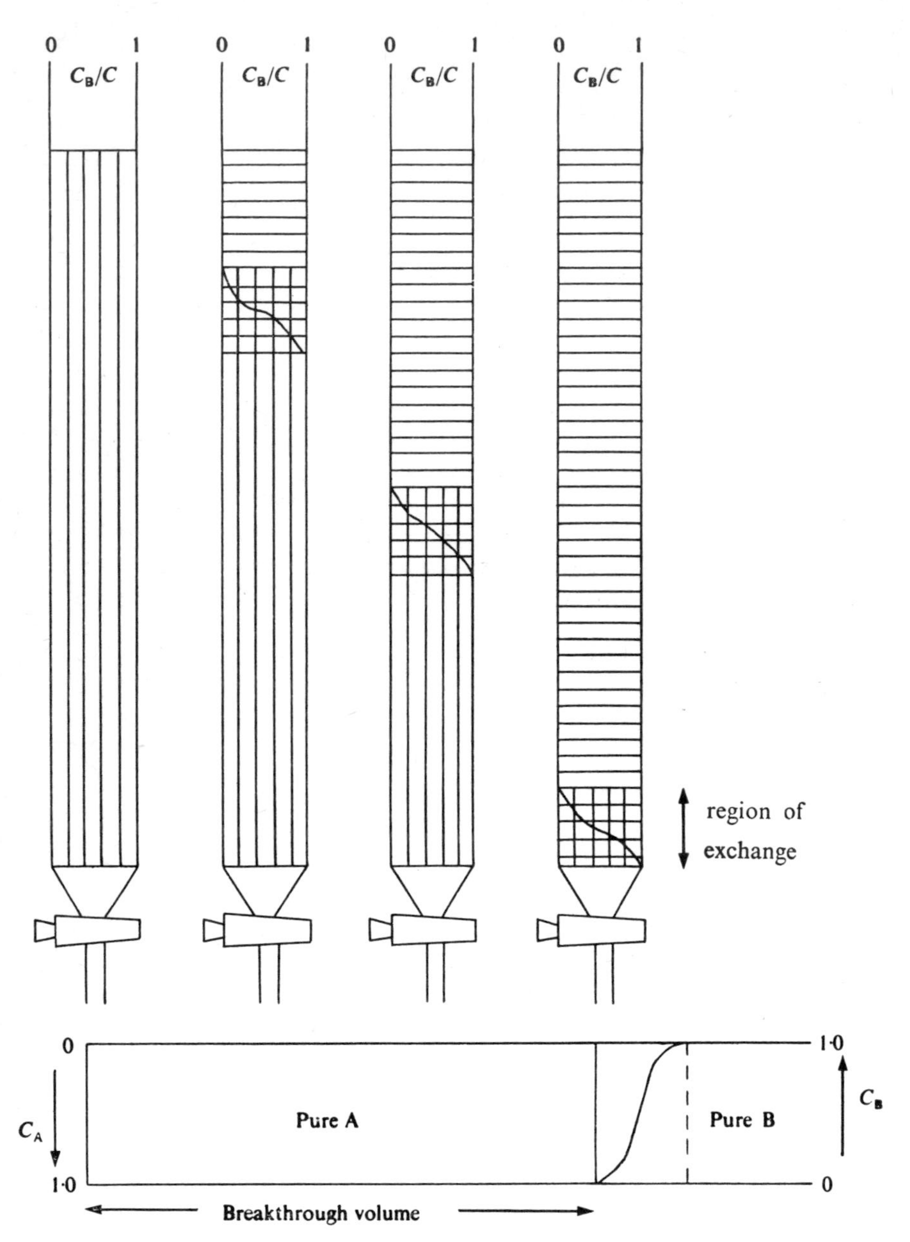

FIG. 6.1. Concentration profiles in column and solution in a displacement of A by B counter ions

of excess acid with a slow flow of distilled water. This process is rather slow, since the acid present in the beads as invading electrolyte must also be removed. When the effluent is no longer acidic the column is ready for use.

To ensure quantitative conversion of a salt MX into the corresponding acid HX, the amount of exchangeable M in the solution should be around one tenth or less of the total capacity of the column bed. A suitable quantity for this column would be 2 mequiv, corresponding to 2 ml of 1N or 20 ml of 0·1N. (Volume in ml multiplied by normality equals the number of mequiv present.)

The water level in the column is drained to just above the resin bed and the sample of solution pipetted carefully onto the top. The solution is allowed to percolate slowly into the column and is washed through at a slow rate with distilled water. In the ion exchange process

$$M^{z+} + |z|\overline{H—R} \rightarrow \overline{M—R_{|z|}} + |z|H^+$$

the solution is converted into HX. The effluent may be collected in a titration flask until no further acid is washed from the column. The flask now contains only HX, which may then be titrated to determine the original concentration of M or, if a mixture of cations was originally present, the total cation content. This is a convenient method of analysis for salt solutions where no obviously simple method of analysis is possible, e.g. sodium perchlorate. If carefully done, this method is reproducible to ±0·2%, and may be used for anions or cations or for the preparation of a sample of any salt, acid or base.

6-2. WATER SOFTENING

If a salt solution is passed through two columns in series, the first a cation exchanger in the hydrogen form and the second an anion exchanger in the hydroxyl form, the final product will be water. In the first column the salt MX will be converted to the acid HX and in the second X is replaced by hydroxyl ion, hydrogen and hydroxyl ions combining to form water.

This process is improved by mixing the exchangers into a single column to give a mixed bed. Under these conditions the exchanging ions M and X are never in the presence of excess of hydroxyl or hydrogen ions, which would reduce their uptake on the column. It is of interest to note that this process also removes dissolved carbon dioxide by anion exchange of the bicarbonate ion,

$$CO_{2(aq)} + H_2O \rightleftharpoons H^+ + HCO_3^- \quad \text{(normal equilibrium saturation)}$$

$$H^+ + HCO_3^- + \overline{R—OH} \rightarrow H_2O + \overline{R—HCO_3}$$

Two points to be noticed are that the anion exchangers are usually less dense and have capacity around 1 mequiv/ml of wet resin. On a volume basis, the mixed bed will have a 2:1 ratio of anion to cation exchanger. Considerable care must also be taken to prevent selective layering in the filling process due to the anion exchanger tending to settle more slowly than the heavier cationic resin. The conductivity of such deionised water is usually comparable or rather better than ordinary distilled water but it is important to note that non-electrolytes, such as sucrose or other non-ionic contaminants, will pass unchanged into the effluent.

6-3. SELECTIVE DISPLACEMENT

Separation of two counter ions B and C by selective displacement involves the use of an ion of intermediate selectivity, A. The selectivity sequence is B > A > C. The ions are loaded on to a column in the A-form forming a narrow layer at the top. A solution of AY will displace C, while B, being more strongly held than A, remains in the column to be displaced by a more efficient feed solution. Kraus[1] has used this technique with great success to separate cations on anion exchangers by their abilities to form complexes with the chloride ions of hydrochloric acid. Cations forming anionic complexes are taken up by the anion exchanger, while those which are uncomplexed or do not form negative complexes are excluded by the Donnan effect (section 3.2).

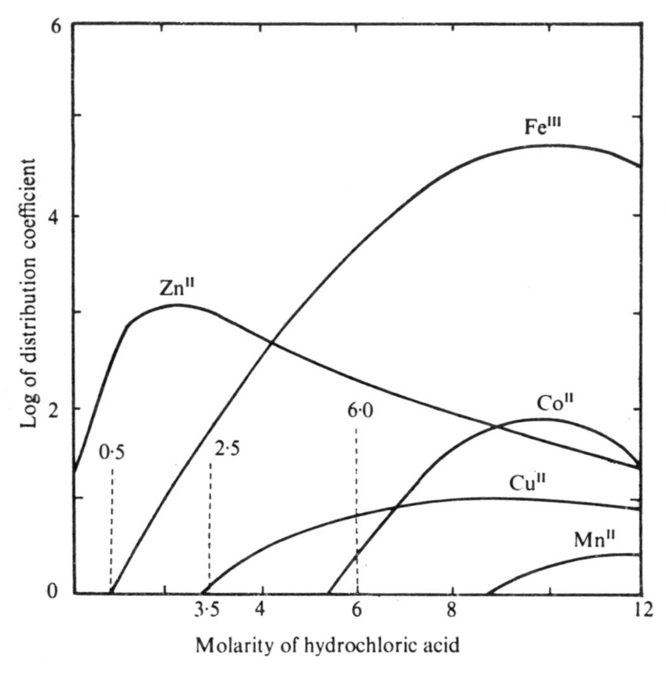

FIG. 6.2. Selective displacement. The profiles of distribution coefficients for some transition metal ions as a function of concentration of hydrochloric acid. Nickel, which forms no complexes with chloride, has a distribution coefficient of zero over the range (taken from reference 2, with permission)

It has been shown that at trace loading the distribution coefficient is a function of the concentration of the developing solution (section 3.5). If complexing occurs, this distribution coefficient will be strongly affected by the degree to which this complexing occurs and the types and charges of the most common complex species. The distribution coefficients of the first row transitional metal ions from Mn^{II} to Zn^{II} are shown in Fig. 6.2 as a function of the concentration of hydrochloric acid in the range zero to 12M.[2] Ni^{II} is the only ion of the series which forms no complexes with the acid in this range of concentrations. If a sample of these ions is placed on to a column of Dowex 1 (anion exchanger) which has been washed with 12M hydrochloric acid, only Ni^{II} will not be absorbed. Development with 12M hydrochloric acid will therefore remove nickel quantitatively, leaving the other ions on the column. At 6M all but manganese ions are retained, and so by progressively reducing the concentration of the HCl solution it is possible to selectively remove each ion in sequence. The method is aptly called selective displacement. Kraus and Moore[3] have completed this experiment and their results are shown in Fig. 6.3.

It is obvious from the logarithmic ordinate of Fig. 6.3 that anion exchangers may have enormous (and as yet unexplained) selectivity for complex ions while the corresponding selectivities for the aquo ions on the normal strong acid exchangers are in no way spectacular. Hydrofluoric acid has also been used in this manner. Both

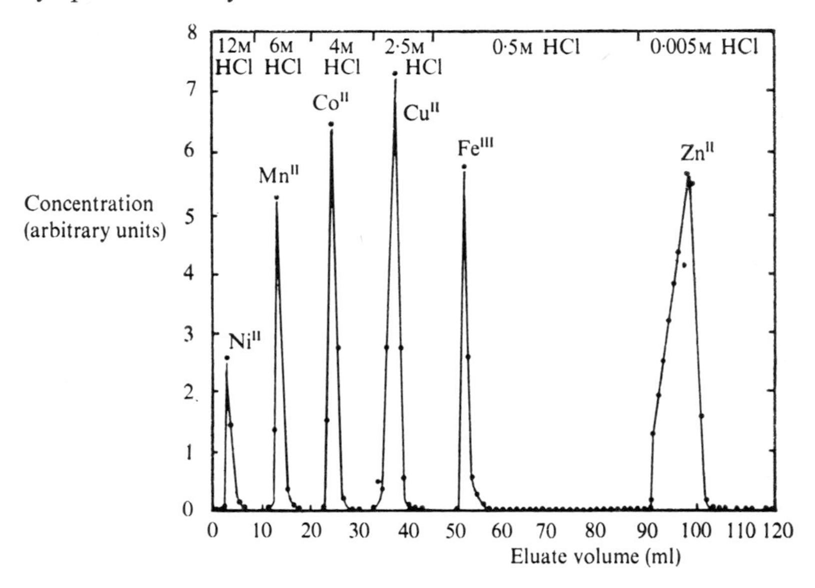

FIG. 6.3. Separation of transition metal ions from Mn^{II} to Zn^{II} by selective displacement with varying concentrations of hydrochloric acid. The column 6·1 × 260 mm contained Dowex 1, a strong base anion exchanger. (Taken from reference 3, with permission.) The concentrations of hydrochloric acid were selected from a knowledge of the distribution coefficients illustrated in Fig. 6.2

acids have the advantage that they can easily be removed from effluent samples by subsequent evaporation.

6-4. ION EXCLUSION AND ION RETARDATION

These are two methods used to separate ionic from non-ionic constituents of a solution. Ion exclusion uses the Donnan exclusion principle (section 3.3). When a mixture of electrolytes, MX_1, MX_2 and non-electrolyte are passed down a column in the M-form, the salts will not be absorbed by the exchanger (especially so if the solution is dilute and Donnan exclusion is almost complete). The non-electrolyte is unaffected by Donnan potentials and is free to distribute itself between the aqueous external and pore solutions. It is therefore more strongly retained under suitable conditions. The degree of uptake will depend strongly upon the particular non-electrolyte present; simple alcohols and acetone have fairly high distribution coefficients while sugars are poorly retained.

If such a mixture is loaded on to a column and developed with water, first the ionic and then the non-electrolyte constituents appear in the effluent. Wheaton and Bauman[4] have used this method to separate a mixture of hydrochloric and acetic acids on a sulphonic acid resin in the hydrogen form. In this case the acetic acid behaves essentially as a non-electrolyte. The common ion effect of the hydrogen ion of the hydrochloric acid suppresses its rather weak tendency to dissociate.

Ion retardation is essentially the opposite effect. The exchanger used has a strong tendency to absorb salt molecules from solution. The exchanger is a 'snake-cage' polyelectrolyte and contains both cationic and anionic fixed charges. These fixed charges normally compensate one another. In the presence of salt however, they will be compensated to electroneutrality by binding both the cations and anions of the salt as 'counter ions'. The non-electrolyte is by comparison rather poorly held, and so by development with water, the non-electrolyte is removed, followed subsequently by the salt. An example is given by Hatch, Dillon and Smith,[5] who separated sodium chloride and glycerol by this method.

6-5. CHROMATOGRAPHY

Chromatography is a technique which allows separation of a mixture according to the different rates of migration of the constituents. It is not confined to any one medium. Conventional chromatography, originally developed as a tool for the separation of organic compounds on alumina and activated charcoal, has been developed to include paper, gas, partition and ion exchange chromatographies. The basis principles and theories are the same in all cases. The most important chromatographic techniques are elution and displacement developments. In both cases, a mixture of the ions to be separated is introduced on to the top of a column and 'developed' with a suitable solution. If the ion of the developing solution has a higher affinity for the column than the mixture, the process is a displacement – the preferred ion of the feed displace all others before it down the column. If the developing ion has a lower affinity for the resin than the mixture, the process is elution – the feed ions pass over the mixture and cause a separation by allowing the components to move down the column at different rates.

The common theories of chromatography are based upon the assumption that local equilibrium conditions are achieved between solute and exchanger at all times. The optimum practical conditions for column use will occur when the rate of ion exchange is high. High rates of ion exchange are favoured by having exchangers with small particle size, since the rate of a particle diffusion controlled exchange is inversely proportional to the square of the radius of the beads (section 4.2). Low cross-linking and high temperatures increase diffusion rates in the bead and so increase exchange rates. These conditions must be balanced by the rather lower flow rates possible when the exchanger particles are very small and the adverse effect of reduced cross-linking upon selectivity (Fig. 3.6).

It is of interest to note that, if the rate is controlled by film diffusion and is therefore a function of the thickness of the unstirred film around the exchanging bead, reduction in flow rate will increase this thickness and reduce the rate of uptake and hence the approach to equilibrium conditions [section 4.3 and Eqn. (4.8)]. The theoretical concentration profiles depend upon an even flow of feed solution through the bed with a front normal to the column length. Uneven particle sizes cause channelling and uneven flow through the bed result in rather more diffuse bands.

6-5a. Displacement development

If two ions A and B are to be separated by displacement development, the exchanger is first converted into the C-form, where C is more poorly selected than either A or B. (The selectivity sequence is A > B > C.) The mixture A, B is loaded on to the column and efficiently displaces C from the top layer of the column. A and B are now held in a narrow band. If this band is now developed by a solution of DY, in which the counter ion, D, is more strongly preferred than A or B, the band A, B will be displaced with a self-sharpening boundary and will, in turn, displace C down the column. As the mixture A–B is displaced, A, having larger distribution on the column, will move more slowly than B and thus as the displacement proceeds the mixture A–B will

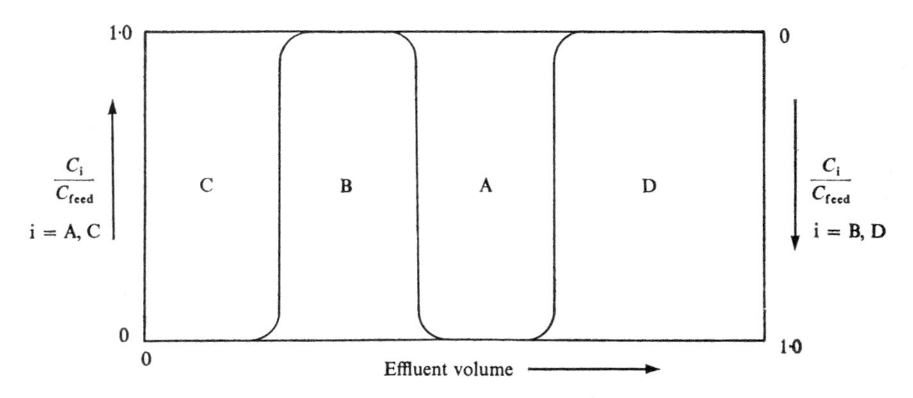

FIG. 6.4. Concentration profiles in the effluent in a displacement development in which the selectivity sequence for the ions is D > A > B > C

separate into two adjacent bands of pure A and pure B. These bands migrate down the column at the same rate; a rate which is now dependent primarily upon the column flow rate. Since each ion displaces the others below it, the boundaries are self-sharpening and do not spread as the development continues. In the effluent the ions appear an increasing order of selectivity C, B, A, D (Fig. 6.4).

The components B and A come off the column as bands with little overlap between them. Since relatively large quantities of ions may be separated in this way, displacement development is primarily suited to preparative techniques, where large quantities of pure samples are required.

A classic example of this method is the separation of the lanthanide elements. The lanthanide elements are formed as the 4*f* atomic orbital is filled. Chemically the ions are almost identical, and their separation by techniques of selective crystallisation was laborious and incomplete. Selectivities on cation exchangers are in the order of La^{3+} to Lu^{3+} across the series. The largest ion having the smallest hydrated radius (La^{3+}) is preferred (section 3.6b), but the differences in selectivity between one ion and another are not sufficient to allow separation and so selective complexing is necessary. Spedding and Powell[6] used a citrate buffer to achieve a separation of the end members of this series. In the ammonium citrate buffer (pH 8) the citrate ion is capable of complexing the lanthanide cations, forming stronger complexes with the ions as their ionic radius decreases down the series. The smaller ions, which are least preferred on the resin, being more strongly complexed, are more strongly retained in solution. The result is that the distribution coefficients of all the ions are reduced, but differences between adjacent ions in the series are increased to a degree which allows separation (section 3.5). The reduction in distribution coefficients is such that

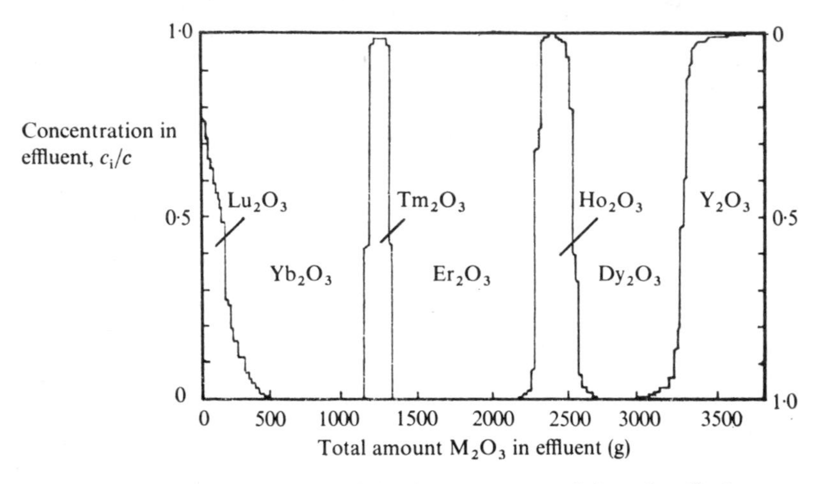

FIG. 6.5. Plant scale separation of the heavy rare earth ions by displacement development. Displacement solution; citrate buffer of pH 8·0 prepared from 0·1 per cent wt citric acid and concentrated ammonia. Resin Nalcite HCR in the hydrogen form. (Taken from reference 8, with permission)

the ammonium ion of the citrate buffer is sufficiently preferred for a displacement to occur (*cf.* electroselectivity, section 3.3). The results of this experiment are shown in Fig. 6.5.

6-5b. Elution development

If two ions A and B are to be separated by elution with ion D, the selectivity sequence must be $D < A < B$. A sample of solution containing A and B is loaded on to the column in the D-form and developed with D ion. Because D is less strongly held, the distribution coefficients λ_A and λ_B are large in the eluting solution (section 3.5), and since the rate of migration of these ions is inversely proportional to λ_A and λ_B (and λ_B is greater than λ_A), there is a progressive separation of A and B as the band moves down the column. Separation is therefore improved by using a longer column, but at the same time the spread of the bands increases. A diagrammatic representation of this separation is shown in Fig. 6.9.

Since elution development requires that the loading of the column be small, it is primarily used in analytical procedures. One of its most spectacular applications was the isolation and identification of the transuranium elements. The naturally occurring elements number 92, the heaviest being uranium (atomic number 92). With the advent of advanced nuclear techniques a series of radioactive isotopes of transuranium elements have been prepared, with atomic numbers up to 102. These elements are obtained in trace quantities – often only a few atoms of any one element may be produced and the matter is further complicated by their very short half-lives, which may be of the order of a few hours or minutes. The original separation of these elements was reported by Harvey, Choppin and Seaborg[7] who used an elution technique on Dowex 50 (sulphonic acid exchanger). The eluting agents were either ammonium citrate or ammonium lactate buffers at an operating temperature of 87°C. In a later publication, Choppin, Harvey and Thompson[8] found that 0·4M ammonium α-hydroxy-isobutyrate was a superior eluant. Their results are shown in Fig. 6.6.

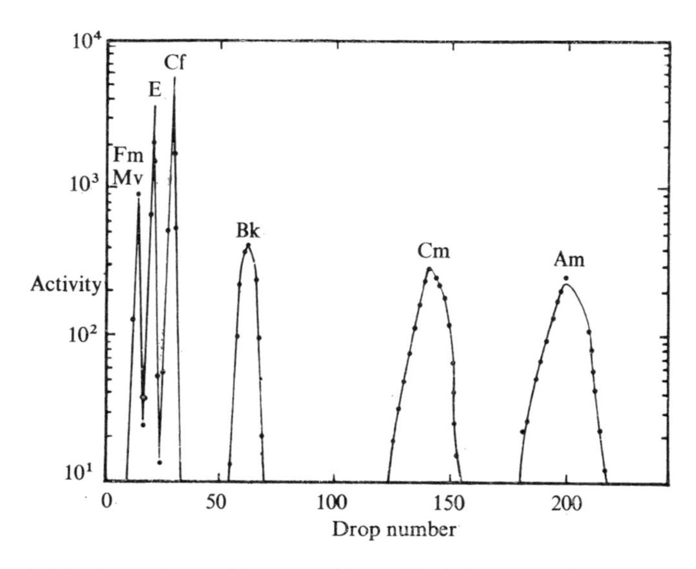

FIG. 6.6. Elution of tripositive actinides with 0·4M ammonium α-hydroxyiso-butyrate. Column: 2 × 50 mm containing Dowex 50 X-12 cation exchanger. Temperature 87°C. (Taken from reference 7, with permission)

Seaborg and Katz[9] acknowledge the near impossibility of characterising these new members of the actinide elements without the rapid and highly selective methods of ion exchange.

6-5c. Complexing anions of weak acids

In the examples given above, complexing agents have been used both in selective displacement and displacement/elution development techniques as a method of altering the relative selectivities of the ions to be separated. When the complexing agent is the anion of a weak acid (citrate, lactate, and α-hydroxyisobutyrate are examples mentioned above), it is obvious that the separation may be either an elution or a displacement (Figs. 6.5 and 6.6).

The normal equilibrium between the undissociated acid and its complexing anions is pH dependent:

$$HX \rightleftharpoons H^+ + X^- \quad \text{monobasic}$$

$$H_2X \rightleftharpoons 2H^+ + X^{2-} \quad \text{dibasic}$$

The greater the acidity (and hence the lower the pH) the fewer ligands are available, and so the degree of complexing is reduced. Below a certain critical level of pH, separations with a given anion will tend to be by elution, since the distribution coefficients of the ions are not sufficiently reduced by complex formation in the solution. Above that pH, i.e. in more alkaline solution, the same feed solution may cause a displacement. The distribution coefficients are so reduced by the increased concentration of ligand that the feed ion is preferred. Helfferich[10,11] has drawn attention to this with the particularly apt examples of two separations of the lanthanide elements in

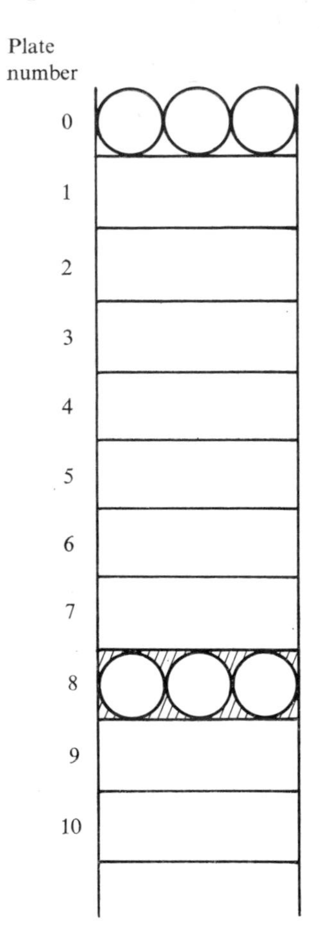

FIG. 6.7. Diagrammatic representation of a column divided into theoretical plates.

citrate buffer solutions. In the first[12] an ammonium citrate buffer of pH 7·5–8·0 causes a displacement. In the second,[13] the citrate buffer contains less ammonia (although more citric acid) and has a pH of 2·5–3·0. The result is that there are fewer citrate ions for complexing and the separation is by elution.

6-6. ELUTION AND THE THEORETICAL PLATE CONCEPT

In elution, the ions to be separated are present as a small proportion of the column capacity. The eluting ion, B, is the original component of the exchanger and the sole counter ion of the feed solution.

In the original theory of elution, Mayer and Tompkins[14] considered the column to be made up of a large number of theoretical plates. Each theoretical plate corresponded to a region in which equilibrium between solution and exchanger was considered complete. The column process was considered to be made up of a series of equilibrium steps. On loading the column with the ion to be eluted (A), it was considered to be equilibrated with the first plate only (theoretical plate 0, Fig. 6.7). The distribution of A between solution and volume Δv_s and the resin volume, Δv_r, in this plate is defined as D_A [Eqn. (6.2)].

$$D_A = \frac{\text{amount of A in the resin}}{\text{amount of A in the solution}} = \frac{\bar{n}_A}{n_A} \text{ (moles)}$$

$$= \frac{\overline{M}_A \Delta v_r}{M_A \Delta v_s}$$

$$= \lambda_A \cdot \frac{\Delta v_r}{\Delta v_s} \tag{6.2}$$

λ_A is the molar distribution coefficient (section 3.5) and M the molar concentration.

D_A is therefore proportional to the molar distribution coefficient. (The value of $\Delta v_r/\Delta v_s$ for an exchanger 200–400 mesh, 0·074–0·037 mm diam., is 1·5.) Since A is a minor component of solution and resin, λ_A will depend upon K_A^B and the concentration of B in the eluting solution (section 3.5).

In this first equilibrium, Δv_s of A solution is equilibrated with Δv_r ml of the resin in the B-form. At equilibrium the fraction of A in the resin is $\bar{n}_A/(\bar{n}_A + n_A) = D/(1 + D)$ and the remainder, $1/(1 + D)$, is in solution. After equilibration with plate 0 the original solution passes down into plate 1 and a fresh solution of B ions enters plate 0. Two separate equilibrations occur and the final equilibrium distribution in the two plates is now

Plate 0

$$\text{resin} \quad \frac{D}{1+D} \cdot \frac{D}{1+D} = \frac{D^2}{(1+D)^2}$$

$$\text{solution} \quad \frac{1}{1+D} \cdot \frac{D}{1+D} = \frac{D}{(1+D)^2}$$

Plate 1

$$\text{resin} \quad \frac{D}{1+D} \cdot \frac{1}{1+D} = \frac{D}{(1+D)^2}$$

$$\text{solution} \quad \frac{1}{1+D} \cdot \frac{1}{1+D} = \frac{1}{(1+D)^2}$$

This process is repeated many times over, and each time a volume Δv_s enters the column and a new series of equilibrations occur, in which A is gradually displaced down the column. The basic column operation is therefore regarded as a series of discrete equilibrium steps rather than a continuous process (as it is in practice).

To illustrate this approach more clearly let us consider the progressive redistribution of A when the distribution coefficient $D = 1$. In Fig. 6.8, the total amount of A in each theoretical plate is calculated as a function of the number of volume elements Δv_s of solution which have entered the column. (For clarity every second profile is plotted.) It is obvious that the distribution of A changes as the elution proceeds, approaching the classic bell-shaped (Gaussian) curve as the number of equilibrations increases. It is also obvious that there is also an increasing spread of A down the

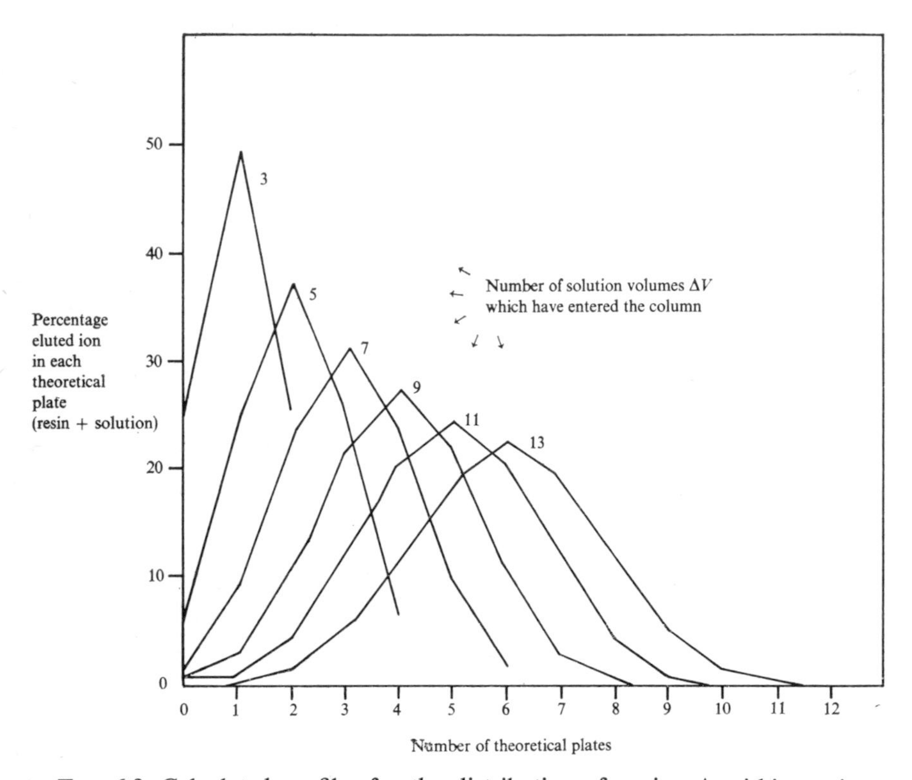

FIG. 6.8. Calculated profiles for the distribution of an ion A within an ion exchange column during an elution experiment. For the purposes of illustration the distribution coefficient, D, is taken as unity

column. If the calculation is continued it becomes obvious that the general expression for the concentration of A in the *solution* at the pth plate after $n \,.\, \Delta v_s$ ml of solution has passed is Eqn. (6.3).

$$C_s(n, p) = \frac{(n + p - 1)!D^{n-1}}{(n - 1)!p!(1 + D)^{n+p}} \tag{6.3}$$

If n and p are very large, as they are if the elution has proceeded for a very large number of steps, it is possible to show that the maximum concentration of solution occurring in plate p_{max} after n_{max} equilibrations will be related to the distribution coefficient D by Eqn. (6.4).

$$n_{max} = p_{max} \,.\, D \tag{6.4}$$

If p_{max} corresponds to the last plate of the column, the next equilibration will give the maximum concentration of A in the eluate. The volume of solution which has to pass the last plate (V) is given by Eqn. (6.5).

$$n_{max} \,.\, \Delta v_S = V \tag{6.5}$$

In this case $p_{max} = N$, the number of theoretical plates on the column, so that $p_{max} \,.\, \Delta v_S = N \,.\, \Delta v_S = v$, the column volume. This is defined as the total volume of solution in the column bed. With these substitutions Eqn. (6.4) becomes Eqn. (6.6).

$$V = v \,.\, D \tag{6.6}$$

Since the column volume is a constant for any given experiment, the elution volume V is proportional to the distribution coefficient, D.

If two ions A_1 and A_2 are originally present in the elution and these are distinguished by different distribution coefficients, D_1 and D_2, the elution peaks will come off the column at eluate volumes V_1 and V_2. The ratio of V_1/V_2 measures the separation of the peaks and is proportional to λ_1/λ_2. It is for this reason that this last ratio is called the separation factor. If the separation factor is unity, then no separation is possible.

From Eqn. (6.6), the difference in the peak elution volumes $V_1 - V_2$ is given by Eqn. (6.7).

$$V_1 - V_2 = v\,(D_1 - D_2) = N \,.\, \Delta v_S(\lambda_1/\lambda_2 - 1) \tag{6.7}$$

The separation of peaks may be increased by increasing the number of plates on the column (Δv_S and λ_1/λ_2 are constants for a column of given dimensions, particle size, and eluting conditions).

This theory holds quite well in practice, but the elution bands are found to be broader and flatter than the theory would predict. The errors lie in the model, which considers the steady flow of feed through the column as a series of discrete equilibrations. Glueckauf[15] has evolved a more accurate theory which is still based upon equilibrium, but in this case considers the feed as a continuous flow and allows for the initial loading to take up more than one plate in the column. The derivation of this approach is beyond the scope of this book but it is useful to consider some of the results.

In place of Eqn. (6.4) we have Eqn. (6.8).

$$\frac{\bar{V}}{X'} = a \tag{6.8}$$

where $\bar{V}$ is the total volume of eluant, which is collected to remove the peak. X' is defined the bed volume used in the elution, $X' = X - \frac{1}{2}X_0$, where X is the total volume of the ion exchanger bed, including solution and X_0 is the volume of the bed occupied by the original loading band. a is a distribution coefficient, which is defined as in Eqn. (6.9).

$$a = \frac{\text{total concentration of the solute per ml of bed}}{\text{concentration of solute in solution}} \tag{6.9}$$

$$= \lambda + \varepsilon$$

λ is the molar distribution coefficient and ε is the fractional void volume of the column i.e. the fraction of the bed volume taken up by solution.

For a chromatographic column containing 200–400 mesh resin ε is equal to 0·39. Once again it is obvious that when two or more ions are to be separated that the elution volumes required will be a function of the separation factor.

The number of theoretical plates in use in the column is given by N' (Eqn. 6.10).

$$N' = 2\pi \left(\frac{c_{max} \,.\, \bar{V}}{m}\right)^2 \qquad (6.10)$$

c_{max} is the maximum concentration of solute in the eluate and $m = \int c \, dv$. N' is the total area under the elution peak and is thus equal to the total amount of this ion in the original sample. Alternatively, N' may be expressed as a function of the band width, β (Fig. 6.9).

$$N' = 8 \left(\frac{\bar{V}}{\beta}\right)^2$$

The band width is defined as $c_{max}/e = 0{\cdot}368 c_{max}$.

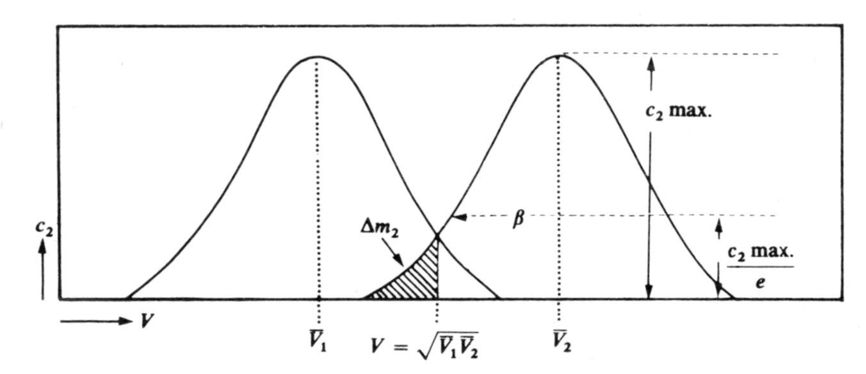

FIG. 6.9. The width and overlap of eluted bands. (Taken from reference 15, with permission)

The number of theoretical plates may therefore be calculated from a single elution experiment. Again it is obvious from Eqn. (6.10) that the peak height, c_{max}, decreases as the square root of the column length for a given loading m. (Thus the band will spread increasingly as the length of the column increases.)

The distribution coefficient, a, may be obtained by passing a solution of solute in eluant through the column until no more solute is retained, that is, until the solution passes through unchanged.

6-7a. Calculation of separation conditions from the Glueckauf theory

If the distribution coefficient is known, the peak elution volume V may be calculated from Eqn. (6.8), but if several solutes are present it is of obvious interest to know the degree of overlap between each band and so the percentage purity of the sample separated. It is assumed that the number of theoretical plates may be taken as a constant for each solute in the separation, and also that the eluate is cut into fractions at the volume, $V = \sqrt{\bar{V}_1 . \bar{V}_2}$. (The geometric mean of the peak volumes) as shown in Fig. 6.9. The impurity content, η, may be calculated as

$$\eta = \frac{\Delta m_2}{m_1 + m_2} \simeq \frac{\Delta m_2}{m_1}$$

Δm_2 is present as impurity in the elution band of solute 1.

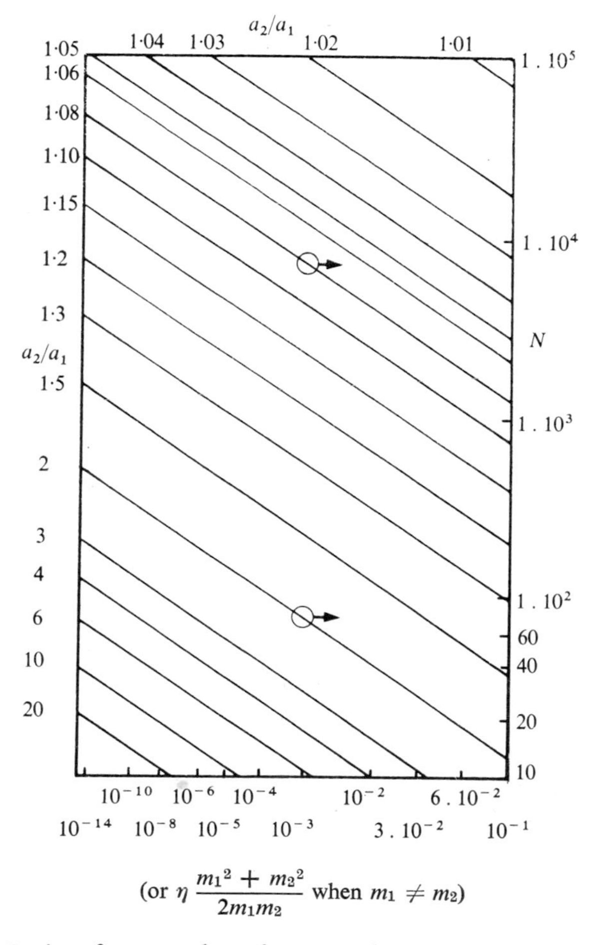

(or $\eta \frac{m_1^2 + m_2^2}{2m_1m_2}$ when $m_1 \neq m_2$)

FIG. 6.10. Purity of separated products as a function of the separation factor (a_2/a_1), number of theoretical plates N and mass ratio m_1/m_2. (Taken from reference 15, with permission)

Glueckauf[15] has produced a convenient chart (Fig. 6.10), which allows the calculation of the required number of theoretical plates to give a predetermined band purity. The abscissa is

$$\eta \text{ or } \eta \left(\frac{m_1^2 + m_2^2}{2m_1m_2}\right) \text{ when } m_1 \neq m_2$$

m_1 and m_2 are the total amounts of solutes 1 and 2 in the separation, $a_2/a_1 = \bar{V}_2/V_1$ is the ratio of the peak elution volumes and $N(\approx N')$ is the number of theoretical plates on the column.

Consider a hypothetical separation in which an impurity content of 0·1% may be tolerated, and equal quantities m_1 and m_2 of solutes 1 and 2 are to be separated. If the ratio $a_1/a_2 = 2$, only 80 theoretical plates are required, but if the ratio is only 1·1, a similar separation would require a column of 8000 plates and would be almost impossible in practice.

6-7b. The height of the theoretical plate

This height decreases with decreasing particle size and under ideal conditions would be approximately equal to the diameter of the resin beads themselves. Values of the order of three diameters would be regarded as rather good in practice since a degree of channelling and uneven flow is unavoidable. Non-equilibrium, caused by slow exchange kinetics or rapid flow rates, has also the effect of increasing the height of the plate.

It is, however, important to note that the height of the theoretical plate is not a constant property of a column, even if the situation is entirely ideal and equilibrium is reached at all times. With particle diffusion control kinetics (section 4.2), the height of the theoretical plate is proportional to the distribution coefficient a of the species. This means that different values of N' apply to different species on the same column and in the same experiment.

REFERENCES

1. K. A. Kraus and F. Nelson, *Proc. Intern. Conf. Peaceful Uses Atomic Energy, Geneva* **7,** 113 (1956).
2. K. A. Kraus and F. Nelson, 'Anion Exchange Studies of Metal Complexes', in *The Structure of Electrolyte Solutions* (W. J. Hamer, Ed.), John Wiley and Sons, New York, 1959, p. 340.
3. K. A. Kraus and G. E. Moore, *J. Amer. Chem. Soc.* **75,** 1460 (1953).
4. R. M. Wheaton and W. C. Bauman, *Ind. Eng. Chem.* **45,** 228 (1953); *Ann. N.Y. Acad. Sci.* **57,** 159 (1953).
5. M. J. Hatch, J. A. Dillon and H. B. Smith, *Ind. Eng. Chem.* **49,** 1812 (1957).
6. F. H. Spedding and J. E. Powell in *Ion Exchange Technology* (F. C. Nachod and J. Schubert, Eds.), Academic Press, New York, 1956.
7. B. G. Harvey, G. R. Choppin and G. T. Seaborg, *J. Amer. Chem. Soc.* **76,** 6229 (1954).
8. G. R. Choppin, B. G. Harvey and S. G. Thomson, *J. Inorg. Nucl. Chem.* **2,** 66 (1956).
9. J. A. Katz and G. T. Seaborg, *The Chemistry of the Actinide Elements*, Methuen, London, 1957.
10. F. Helfferich, *Ion Exchange*, McGraw-Hill, New York, 1962.
11. F. Helfferich, *Advances in Chromatography*, Vol. 1 (J. C. Giddings and R. A. Keller, Eds.), Edward Arnold, London, 1965, Chapter 1.
12. F. H. Spedding, E. I. Fulmer, J. E. Powell, T. A. Butler and I. S. Yaffe, *J. Amer. Chem. Soc.* **73,** 4840 (1951).
13. F. H. Spedding, A. F. Voigt, E. M. Gladrow and N. R. Sleight, *J. Amer. Chem. Soc.* **69,** 2777 (1947). F. H. Spedding, A. F. Voigt, E. M. Gladrow, N. R. Sleight, J. E. Powell, J. M. Wright, T. A. Butler and P. Figard, *J. Amer. Chem. Soc.* **69,** 2786 (1947).
14. S. W. Mayer and E. R. Tompkins, *J. Amer. Chem. Soc.* **69,** 2866 (1947).
15. E. Glueckauf, *Trans. Faraday Soc.* **51,** 34 (1955).

FURTHER READING

E. Heftmann (Ed.), *Chromatography*, Reinhold, New York, 1967.
J. C. Giddings and R. A. Keller (Eds.), *Advances in Chromatography*, Vol. 1, Edward Arnold, London, 1965.
O. Samuelson, *Ion Exchange Separations in Analytical Chemistry*, John Wiley and Sons, New York, 1963.
F. Helfferich, *Ion Exchange*, McGraw-Hill, New York, 1962, Chapter 9.

7 Inorganic ion exchangers

Inorganic ion exchangers are both the oldest and the newest ion exchange materials. Natural and synthetic aluminosilicates were the first ion exchangers, and the first commercial synthetic ion exchanger of this type was produced by Harm and Rumpler[1] in 1903. Such 'permutits' were in common use until the advent of synthetic organic resin exchangers and thereafter have disappeared from common usage. In more recent times pure crystalline zeolites, which had been previously obtained from natural mineral deposits, were synthesised by Barrer and his co-workers. As exchangers they have some remarkable selectivities, and certain of them, because of their regular and restricted pore sizes, will exclude ions on a basis of size alone. Their use in common ion exchange processes is severely restricted by their excessively slow exchange rates, which are on average some thousand times slower than common resin exchangers. As a result their major place in current techniques is as molecular sieves, absorbing small molecules and excluding those which are larger than the crystal pores. Several of these sieves are prepared by the Linde Corporation with a variety of pore sizes ranging from 3 to 13 Å.

In the early 1950's a series of new inorganic ion exchangers was developed, primarily as tools for the separation of ionic fission products. The aim was to produce exchangers combining good chemical and thermal stabilities with the ability to withstand radiation levels which would destroy conventional resin exchangers. At the same time it was hoped to discover materials with improved selectivities.

To some degree these aims have been successfully achieved, and in the remainder of this chapter the general physical chemistry and more important separations are discussed. Preparation and structural chemistry are considered in Chapter 2.

7-1. ALUMINOSILICATE EXCHANGERS

These exchangers, commonly known as zeolites, are essentially the only truly crystalline materials in the study of ion exchange. Many of the more commonly used crystals have been studied by X-ray diffraction techniques and their structures solved. Crystal pore sizes and the distribution of charged sites may therefore be specifically stated and used in the interpretation of kinetic and equilibrium studies. The ion exchange properties are dominated by this structural regularity and it is not surprising that they usually show regular and sometimes remarkable selectivities. The capacities of the more common natural zeolites are given in Table 2.1.

Certain extreme selectivities based upon ion sieve effects are possible, and a good example of this is given by analcite, which will freely exchange with rubidium and not caesium ions although they are almost identical chemically. This ultimate selectivity appears to be purely an ion sieve effect, although the ionic radius of Cs^+ is larger by only 0·15 Å. Ultramarine will similarly distinguish between potassium and caesium ions.

The relative exchanging abilities of some common ions are given in Table 7.1.[2] A remarkable example of partial ion sieve effects and site specificity is given by Barrer and Meier[3] in a series of exchanges involving silver, sodium and thallous ions on the synthetic zeolite Linde sieve 4A.

TABLE 7.1. Relative Uptake of Monovalent Cations by Common Zeolites[2]

Exchanger	Extent of Exchange					
	Zero ⟶					Complete
Ultramarine	Cs^+		NH_4^+, Rb^+		Li^+, Na^+, Tl^+	Na^+, Ag^+
Analcite	Cs^+		Li^+	Rb^+, H^+		Na^+, K^+, NH_4^+, Tl^+, Ag^+
Mordenite				Li^+	K^+, $MeNH_3^+$	H^+, Na^+, NH_4^+
Chabazite					Li^+, Cs^+, $MeNH_3^+$	H^+, Na^+, K^+, NH_4^+, Ag^+
Faujasite		NEt_4^+	NMe_4^+ Me_3NH^+		$Me_2NH_2^+$ $MeNH_3^+$	H^+, NH_4^+, Na^+, K^+, Rb^+, Cs^+, Tl^+, Ag^+

There are thirteen ion exchange sites in a unit cell of this material and, of these, twelve are identical and the thirteenth is placed in a structurally unique position. The sites are designated α and β respectively. All sites are available for exchange with sodium and silver ions, but only the twelve α sites for the larger thallous ion (Tl^+). The isotherm for Na/Ag exchange is S-shaped (Fig. 7.1a) and may be considered as the sum of two regular isotherms (Fig. 7.1b). Up to 12/13 conversion the exchanger strongly prefers silver ion and only α sites are involved. In the subsequent conversion of the β sites, the selectivity is reversed to a preference for sodium ion. This results in an inversion of the isotherm over the last 1/13 of the conversion. The exchanger is therefore bi-functional and behaves as a mixture of two exchangers in the proportions 12/13 and 1/13, with quite opposite selectivities. Analysed in this way selectivity coefficients may be calculated.

$$K_{\text{Na}}^{\text{Ag}} = 535 \quad \alpha \text{ sites}$$

and

$$K_{\text{Na}}^{\text{Ag}} = 0{\cdot}135 \quad \beta \text{ sites}$$

The isotherms for Na/Tl and Ag/Tl are regular but involve only the twelve α sites. The selectivity coefficients are constant over the whole range of conversion and are

$$K^{\mathrm{Tl}}_{\mathrm{Na}} = 259 \pm 2 \quad \alpha \text{ sites only}$$

$$K^{\mathrm{Tl}}_{\mathrm{Ag}} = 0{\cdot}046 \quad \alpha \text{ sites only}$$

$$K^{\mathrm{Ag}}_{\mathrm{Na}} = 535 = \frac{[\overline{\mathrm{Ag}^+}][\mathrm{Na}^+]}{[\overline{\mathrm{Na}^+}][\mathrm{Ag}^+]} = \frac{[\overline{\mathrm{Tl}^+}][\mathrm{Na}^+]}{[\overline{\mathrm{Na}^+}][\mathrm{Tl}^+]} \cdot \frac{[\overline{\mathrm{Ag}^+}][\mathrm{Tl}^+]}{[\overline{\mathrm{Tl}^+}][\mathrm{Ag}^+]} = \frac{K^{\mathrm{Tl}}_{\mathrm{Na}}}{K^{\mathrm{Tl}}_{\mathrm{Ag}}} = \frac{25{\cdot}9 \pm 2}{0{\cdot}046} = 562 \pm 46$$

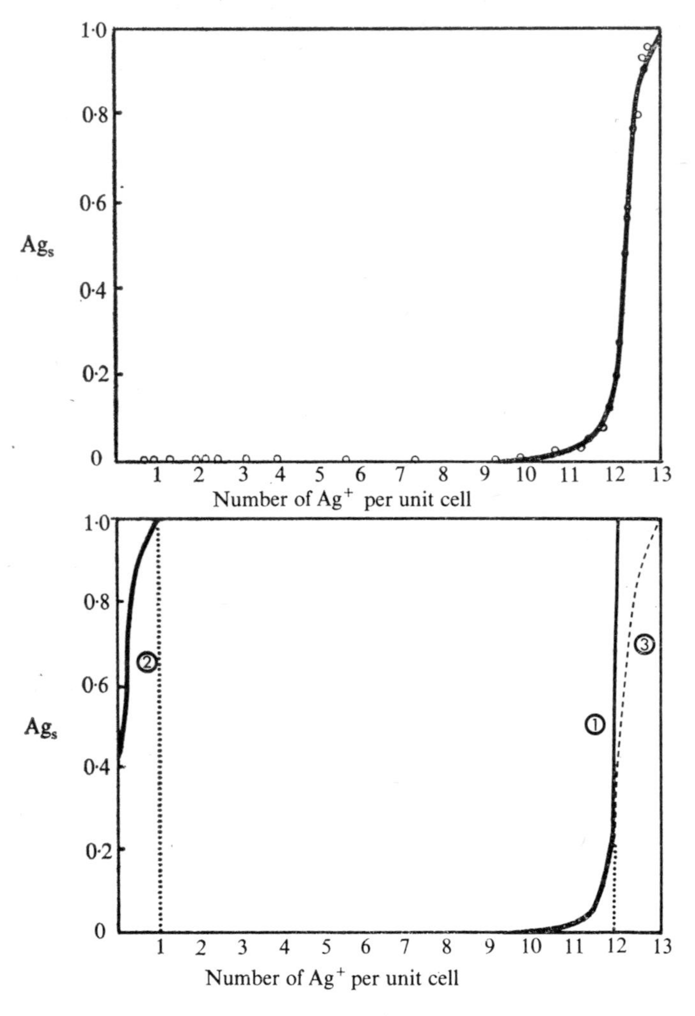

FIG. 7.1. Exchange isotherms for sodium/silver exchange on Linde Sieve 4A.
(a) Sodium–silver exchange (experimental).
(b) An analysis of the experimental isotherm in terms of α and β sites in the exchanger

α sites only : (1)
β sites only : (2)
α and β sites: (3)

(Taken from reference 3, with permission)

This is an example of the triangle rule, showing that, within experimental error, the measurement of two such equilibrations may be used to calculate the third. It must be noted that selectivity coefficients can only be used in this manner if they are regular (invariant over the whole range of conversion). The triangle rule is strictly applicable to thermodynamic equilibrium constants (section 3.4).

S-shaped isotherms may therefore occur in exchangers where there are two quite different ion exchange sites with quite opposite selectivities for the ions being exchanged. This is not the sole criterion as other examples will illustrate. The regular behaviour of the α sites of Linde sieve 4A is also found in a few cases for other zeolites. Basic sodalite, for example, contains only one type of site and shows regular behaviour in exchanges involving Na, K, Li and Ag ions. Selectivity constants for any pair are quite constant and independent of the fractional composition of the exchanger. Such behaviour is quite rare in zeolites and almost unheard of in resin exchangers. Most ion selectivities decrease as the exchanger phase becomes progressively saturated with the selected ion (*cf.* Fig. 3.5).

7-1a. Irregular behaviour in zeolite exchangers

Certain zeolites show selectivity inversion and so S-shaped isotherms simply because the accomodation of the larger ions becomes more difficult as the conversion proceeds. For purely steric reasons the occupancy of adjacent sites by two large ions is made less favourable by their mutual interactions. A more stable state is achieved when adjacent sites are occupied by a large and a small ion or by two small ions. If the large ion is preferred on the matrix, its selectivity will be reduced or even reversed as it becomes

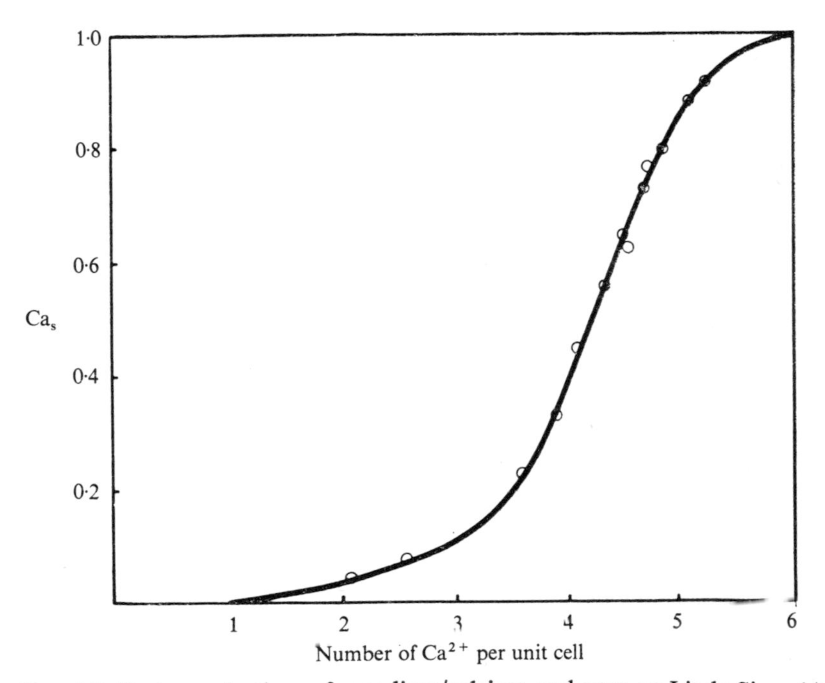

FIG. 7.2. Exchange isotherm for sodium/calcium exchange on Linde Sieve 4A. (Taken from reference 3, with permission)

more probable that it will occupy nearest neighbour sites, that is, as the conversion proceeds. If these effects are large a selectivity inversion may occur. An example is found once more in the exchange on Linde sieve 4A. The α sites behave regularly for exchanges with Ag, Tl and Na ions, but Na–Ca exchange is irregular and selectivity reversal (with the resultant S-shaped isotherm) occurs as the calcium ion becomes a major constituent of the matrix (Fig. 7.2). Examples seem to occur most often when divalent ions are involved.

7-1b. Irregular behaviour due to phase transitions

In certain exchanges the conversion from one ionic form to another involves a rearrangement of the crystalline zeolite lattice. Analcite and Leucite are the sodium and potassium forms of the same mineral, but differ in crystalline form. Under suitable conditions it is possible to convert from one to the other by a simple ion exchange process.

$$K^+ + \underset{\overline{Na}}{\text{Analcite}} \rightleftharpoons \underset{\overline{K}}{\text{Leucite}} + Na^+$$

In this conversion some 20% of the exchange sites on the analcite may be occupied by potassium ions before a spontaneous lattice rearrangement occurs into the leucite form. Similarly, in the reverse sequence as soon as the sodium content of the leucite exceeds 20% a conversion to the analcite lattice is initiated. During the lattice rearrangement both types of crystalline lattice co-exist. Since the conversions occur at different compositions in the forward and reverse exchanges, there is distinct hysteresis in the isotherms; the isotherms for forward and reverse exchanges do not coincide.

7-2. ZIRCONIUM PHOSPHATE

Zirconium phosphate is the most commonly used of all the inorganic ion exchangers. Many preparations appear in the literature (section 2.7) but most of the applications involve the non-crystalline gel material, which has a phosphate to zirconium ratio of between 1·0 and 1·9. These materials differ only qualitatively in selectivity patterns and in particular are distinguished by very high selectivities for caesium ions.

Since the primary aim of most research has been to achieve suitable separations, most of the equilibrium data concerns distribution coefficients. Veselý and Pekárek[4] have produced a table of distribution coefficients determined as a function of composition and temperature (Table 7.2). In this Table the distribution coefficient k_d is defined as the amount of the ion in one gram of the exchanger divided by the solution concentration in mmole/ml and has the dimensions ml/g. K_d and the molar distribution coefficient λ_m are related via the density ρ of the exchanger.

$$\lambda_m = K_d \cdot \rho$$

It is important to distinguish between gel materials and the truly crystalline material prepared by Clearfield,[5] $Zr(HPO_4)_2H_2O$. This material, which we will designate ZrPx, appears to be made up of a layer structure of planes of zirconium atoms, held together by intraplanar phosphate groups, amounting to half the total present. The remaining phosphates are involved in interplanar bonding and are distinguished chemically by having a lower acidity. From X-ray powder diffraction, the interplanar distance between layers of zirconium ions is taken to be 7·6 Å. Confirmation awaits the elucidation of the complete structure by single crystal diffraction studies. The exchange of caesium on ZrPx illustrates the differences between this and the gel forms. The exchange of caesium on ZrPx appears to involve rather less than the total available exchange sites. Initially, only the more acidic half of the monohydrogen phosphate groups are involved, eventually causing the destruction of the zirconium

TABLE 7.2. Distribution Coefficients on Zirconium Phosphate as a Function of Drying Temperature and of PO_4:Zr Ratio[a]

Cation	K_d(ml/g)					
	PO_4:Zr = 1·0			PO_4:Zr = 1·87		
	40°	260°	1000°C	40°	260°	1000°C
Na^+	—	—	—	10·0	8·8	—
K^+	9·5	8·0	9·6	19·5	21·0	3·7
Cs^+	26·3	14·8	10·0	349·0	126·0	4·3
Sr^{2+}	—	—	—	1·0	$<1·0$	—
ZrO^{2+}	361·0	252·0	2·5	1400·0	450·0	24·0
UO_2^{2+}	$<1·0$	—	—	897·0	176·0	—
Ce^{3+}	5·1	2·6	1·6	43·5	3·9	2·7
Fe^{3+}	37·9	34·2	$<1·0$	6570·0	315·0	3·7
Bi^{3+}	700·0	219·0	$<1·0$	$>8000·0$	1000·0	8·7
$RuNO^{3+}$	5·3	—	—	5·8	—	—
Th^{4+}	130·0	75·0	$<1·0$	2760·0	504·0	5·5
Pu^{4+}	1290·0	359·0	16·0	5580·0	684·0	32·4

[a] Medium: 0·1M in nitric acid and 10^{-3}N in cation. Data taken from V. Veselý and V. Pekárek, *J. Inorg. Nucl. Chem.* **25,** 687 (1963).

layers. When these layers break down, some interplanar groups are made available for exchange to caesium and in this way more than half of the phosphate groups are found to exchange with the ion. In ZrP gels all the phosphate groups exchange with caesium and in the Cs-form hydrolysis is diminished. For ZrPx the reverse is true. Both effects seem to be due to the more open and less structured matrix of the gel. Again there are quite different selectivity patterns with ZrPx having a lower affinity for Cs^+ than Na^+, while the gel material has the sequence $Cs^+ > Rb^+ > Na^+$ in acid solutions and the reverse in alkaline.[6]

Since most of the current literature involves gel material, further discussion will be confined to this type of zirconium phosphate. It is, however, useful to remember that such gel material ranges widely in composition and may indeed show a degree of crystallinity without attaining the ultimate structure of ZrPx, with its attendant structurally imposed characteristics. At low pH the hydrolysis of ZrP is suppressed, so it is of obvious interest to use acidic solutions for all work on the gel salts of zirconium or other similar exchangers. Kraus[7] has noted that the exchangers zirconium tungstate, phosphate and molybdate have a remarkable selectivity for the alkali metal ions in acidic solutions. This is particularly so for caesium, which is so strongly absorbed

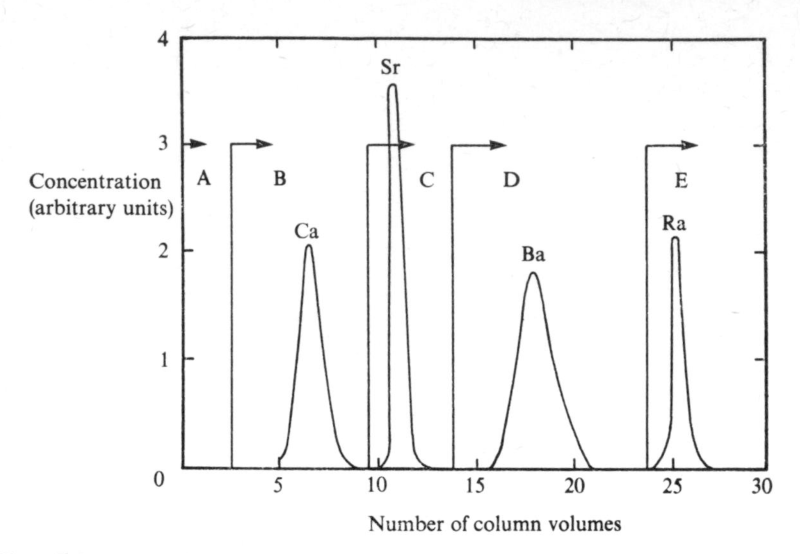

FIG. 7.3. Separation of the alkaline earths at tracer concentrations on zirconium molybdate. (Taken from reference 7, with permission)

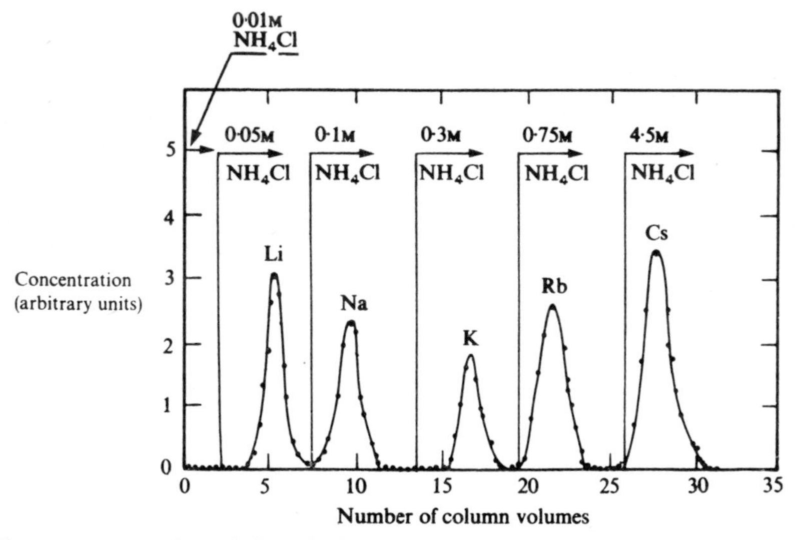

FIG. 7.4. Separation of the alkali metal ions on zirconium tungstate. (Taken from reference 7, with permission)

that it may be isolated, essentially uniquely, from all other elements of the periodic table. Figures 7.3, 7.4 and 7.5 illustrate chromatographic separations of alkali and alkaline earths on a selection of zirconium-based exchangers.[7,8]

Equilibrium studies of the exchanges Cs–H, Cs–Rb, Cs–K and Rb–H have been completed over the whole range of exchanger loadings. Cs–K and Cs–Rb[9] exchanges are quite regular and show the selectivity sequence Cs $>$ Rb $>$ K. It is worth noting that these exchanges occur at ambient pH $\approx$ 3, caused by hydrolysis

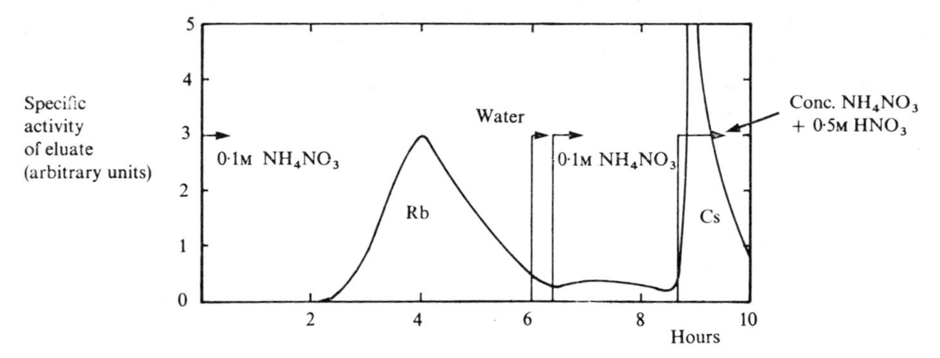

FIG. 7.5. Separation of caesium and rubidium in macro amounts on zirconium phosphate. Temperature 83°C. (Taken from reference 8, with permission)

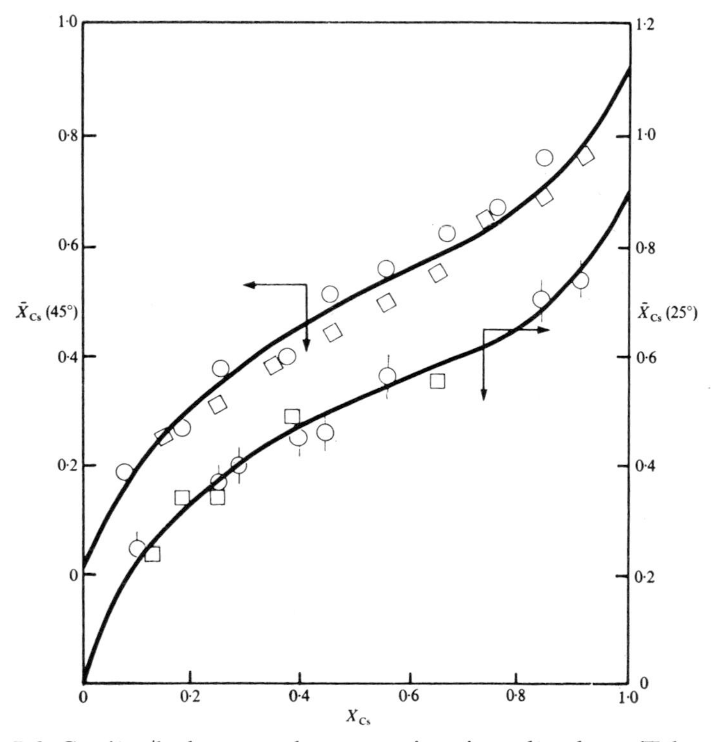

FIG. 7.6. Caesium/hydrogen exchange on zirconium phosphate. (Taken from reference 10, with permission)

of the exchanger in water, and so correspond to the distribution data obtained at trace levels. The exchanges of Cs–H[9,10] and Rb–H[9] are anomalous, showing S-shaped isotherms and (Fig. 7.6) selectivity reversal. Selectivity coefficients for Cs (or Rb) ions are extremely high when the exchanger is essentially in the hydrogen form, and decrease progressively as the proportion of metal ion approaches 50% of the exchanger capacity. Thereafter the selectivity is for the hydrogen ion. The observations of Kraus noted above are therefore substantiated, stressing the importance of trace loadings for the highest selectivities and indicating the difficulties in attempting spectacular separations with macro amounts of alkali metal ions.

As we have seen in section 7.1, such selectivity reversals may be caused by the presence of two exchange sites with quite different selectivity behaviour. In the crystalline material ZrPx, there is evidence for two such sites with different acidic strengths, in the ratio 1:1. If these still persist in the gel materials, the more acidic sites would prefer Cs > (Rb >) H, while the more weakly acidic ones would by definition tend to retain hydrogen ion. Provided these selectivities were sufficiently different, a sharp inversion of selectivity would occur at 50% Cs ion loading, corresponding to the complete conversion of the more strongly acid sites. This behaviour, together with the observed reversal in selectivity for the alkali metal ions in alkaline solution, is similar to that observed by Bregman[11] (section 3.6b) on organic phosphonic acid exchangers. The phosphonic acid group is essentially similar to phosphate and it is tempting to assume that the selectivity behaviour of ZrP has the same basic mechanism. It would, however, be necessary to assume that both acidic hydrogens were placed upon the same phosphate. Much less data is available for other exchangers of the ZrP type, but in general their chemical and physical characteristics are similar. Maeck, Kussy and Rein[12] have obtained the distribution coefficients of all the elements of the periodic table on the exchangers ZrP, Zr tungstate (and zirconia) as a function of pH in the range pH 2 to pH 5. This data should serve as a useful starting point for assessing these materials for any intended separation.

7-3. HETEROPOLYACIDS AND FERROCYANIDES

Heteropolyacids are typified by the behaviour of ammonium molybdophosphate (AMP). Schroeder[13] has determined distribution coefficients for univalent ions on AMP in a medium 0·1M in ammonium nitrate. These are shown in Table 7.3 and illustrate the wide range of selectivity shown towards the alkali metals. In particular, the

TABLE 7.3. Distribution Coefficients of some Univalent Cations upon Ammonium Molybdophosphate[13]

Ion	Na^+	K^+	Rb^+	Cs^+	Ag^+	Tl^+
K_d (ml/g)	0	4	190	5500	26	4300

selectivity for Cs^+ (and Tl^+) ions is so enormous that once absorbed they may be removed only with difficulty by very concentrated acid of saturated ammonium nitrate. Caesium ion is still selectively removed from media 3M in nitric acid and in some cases it is found to be most convenient to remove it by dissolving the exchanger in ammonia.

Experiment shows that di- and trivalent cations are poorly held in strongly acid solutions and must be absorbed at pH's between 2 and 5. Common practice involves the use of sodium acetate buffer (pH 4–5) for the absorption step followed by elution or displacement with more acidic solutions. As noted in Chapter 2, AMP is normally obtained as very fine microcrystals, less than 200 mesh size. Large columns of such fine material would have impossibly slow flow rates and so it has become practice to use asbestos fibre as a support. Micro separations may be obtained using one or two grams of this material in a simple filter bed. Subsequent investigations of Smit[14] allow the preparation of relatively stable conglomerates of crystals which are sufficiently stable for solumn operation. Excellent separations of the alkali metal ions have been achieved using ammonium nitrate as a displacing feed. Sodium ion is quantitatively removed by 0·01M salt and by progressively increasing the concentration each ion in turn may be removed until finally caesium ion is displaced with saturated ammonium nitrate.[15]

The ferrocyanide exchangers are relatively newcomers and little work has been done upon them to date. Kourim, Rais and Million[16] have determined distribution coefficients of the alkali metal ions on zinc ferrocyanide as a function of ammonium nitrate and nitric acid concentrations. In all cases log K_d vs. log $M_{NH_4^+}$ (or log M_{H^+}) (M = molar concentration) are linear and of slope -1, indicating a true ion exchange process as distinct from an adsorption (section 3.5). The distribution coefficients at 0·1M in ammonium nitrate are of similar order to those of AMP quoted in Table 7.3. Efficient separations of the alkali metal ions have been achieved at both trace and milligram amounts. Particle sizes are large enough for conventional column operation and there is apparently no deterioration over a number of cycles of use. This may well indicate that further study may show them to be superior to the heteropolyacid exchangers for common or repetitive use.

Much work remains to be done on the new inorganic exchangers before we may assess their full position in the field of ion exchange. Little is known of their selectivities for complex or organic cations, but what does stand out is their universally high selectivities for the alkali metal ions and here they do fill a gap. Separation factors for the alkali metal ions on PSSA exchangers are very close to unity and separation of the heavier ions is impossible. [In most other cases selective complexing may be used to give enhanced separations (section 6.5), but since the alkali metals are essentially non-complexing this cannot be used.]

Their major disadvantages are due to hydrolysis and in some cases solubility. Those which are gels may vary widely from batch to batch of material prepared in apparently identical methods and so it is wise to conduct all measurements upon the same batch.

7-4. HYDROUS OXIDES

The preparation, structure and general modes of exchange are discussed in Chapter 2. As with the other gel exchangers, their capacities and selectivities are strongly affected by the mode of preparation and particularly by the drying temperature. Kraus[7] has shown that zirconia still retains some ion exchange capacity after drying at 800°C. Measurements of distribution coefficients for bromide–nitrate exchange show the mass action law to be obeyed, and thus indicate that the absorption mechanism is in fact ion exchange (Fig. 3.3). In acid solutions, all but silica gel show anion exchange characteristics. Zirconia, which is probably the most stable chemically, has rather unspectacular selectivities for most common univalent anions (Fig. 7.7), but has been shown to have increasingly high selectivities for di- and trivalent anions. In particular, phosphate and dichromate ions are absorbed chemically and essentially irreversibly.

Possibly the most interesting prospect lies in supported membranes of these materials. One such application has been noted in section 5.9 in the analogue system for active transport. Kraus[17] has recently used colloidal zirconia in hyperfiltration studies. A small proportion of colloidal hydrous zirconia is added to the salt solution which is then pressed through a very fine silver frit. The zirconia partially blocks the frit and establishes it as an anion exchange membrane of relatively high water permeability. Applying pressures of around 35 atmospheres causes water to pass and leaves the bulk of the salt in the original solution. The mechanism is based upon the

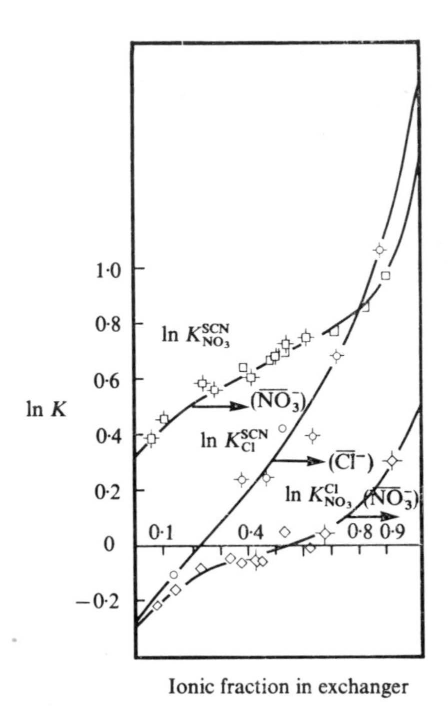

FIG. 7.7. Thiocyanate/nitrate, thiocyanate/chloride and chloride/nitrate exchanges on hydrous zirconia. (Taken from G. H. Naneollas and R. Paterson, *J. Inorg. Nucl. Chem.* **29**, 565 (1967), with permission.)

ion exchange characteristics of zirconia. The solutions are somewhat acid and hence the zirconia behaves as an anion exchanger. The essential ion exchange mechanism is borne out by the observation that salts of di- and trivalent cations are more strongly excluded from the effluent, as would be expected from Donnan considerations. (These are the co-ions of the system.)

7-5. KINETICS OF ION EXCHANGE UPON INORGANIC EXCHANGERS

Kinetic studies are of fundamental importance in deciding the possible use of an exchanger. Zeolites are undoubtedly among the most interesting from equilibrium studies. The combination of ion sieve and specific site selectivities would render them the most powerful tools in ion exchange, if it were not for their excessively slow rates of ion exchange. Column operation demands a relatively rapid ion exchange process in order that the effective theoretical plate may be of minimal dimensions, and even the simplest batch equilibration would be rather impractical if it required some days for completion. Barrer[2] has studied the kinetics of ion exchange on a number of zeolites. Typical diffusion coefficients are around 1×10^{-13} cm²/sec at a temperature of 95°C and some hundred times smaller at room temperature. At this value exchanger diffusion rates are some hundred million times smaller than free ionic diffusion in solution. Half-times of exchange would be of the order of days or weeks compared with minutes or seconds for resin exchangers. It is for this reason that the zeolites have essentially no column operations in ion exchange and are seldom used except for highly specialised or theoretical work.

Nancollas and Paterson[18] and Amphlett and McDonald[19] have obtained quantitative results of some simple ion exchange processes on zirconium phosphate and hydrous thoria, which all show normal diffusion controlled kinetics (Chapter 4). Table 7.4 summarises their results.

TABLE 7.4. Kinetic Data for Simple Ion Exchange Processes on Zirconium Phosphate and Hydrous Thoria[18,19]

Exchanger	Exchange	Particle Diffusion Coefficients $\bar{D}$ (cm²/sec)	Temperature (°C)	pH	Reference
ZrP (gel)	Cs^+	$1{\cdot}1 \times 10^{-8}$	20	2·5	19
	NMe_4^+	$1{\cdot}7 \times 10^{-9}$	—	—	
ZrP*[a]	Na^+–H^+	1×10^{-6}	25	3·0	18
$ThO_2 \cdot nH_2O$	Na^+–H^+	4×10^{-7}	25	13·0	18
		15×10^{-7}			
$ZrO_2 \cdot nH_2O$	Cl^- Self-diffusion	4×10^{-7}	25	1·0	20
Carboxylic Resin Amberlite IRC 50	Na^+–H^+	$3{\cdot}9 \times 10^{-9}$	25	—	21
Sulphonic Acid Resin		Generally around 1×10^{-6}	—	—	

[a] This zirconium phosphate sample had some crystalline properties and the exchanger particles consisted of conglomerates of more finely divided material.

The calculation of $\bar{D}$ involved non-spherical particles and the knowledge of the mean effective radius was obtained either by direct particle size measurements using a calibrated microscope, or from surface area measurements obtained from measurements of surface adsorption of dye. The latter were found to be more effective. The general treatment of results followed the simple isotope theory for particle diffusion† (section 4.2). Exchange rates were found to be inversely proportional to the square of the effective radius, and independent of stirring and external solution concentration in the range 0·01 to 1·0M – further confirmation of particle diffusion as the rate

† It is a curiosity that the mutual diffusion coefficient for Na^+–H^+ exchange is a constant over a wide range of exchanger conversion.

controlling step. In general the diffusion coefficients are intermediate between those of sulphonic acid and carboxylic resin exchangers. The larger diffusion coefficient observed for ZrP* is undoubtedly due to the fact that the exchange diffusion process involves passage within the aqueous channel of the aggregate particle, as well as within the constituent micro-particles, reducing the overall diffusion coefficient.

REFERENCES

1. F. Harm and A. Rumpler, *5. Intern. Cong. Pure Appl. Chem.* **1903,** 59.
2. R. M. Barrer, *Proc. Chem. Soc.* **1958,** 99; *Chem. Ind. London* **1962,** 1258.
3. R. M. Barrer and W. M. Meier, *Trans. Faraday Soc.* **55,** 130 (1959).
4. V. Veselý and V. Pekárek, *J. Inorg. Nucl. Chem.* **25,** 697 (1963).
5. A. Clearfield and J. A. Stynes, *J. Inorg. Nucl. Chem.* **26,** 117 (1964).
6. S. Ahrland and J. Albertsson, *Acta Chem. Scand.* **18,** 1861 (1964).
7. K. A. Kraus, H. O. Phillips, T. A. Carlson and J. S. Johnson, *Intern. Conf. Peaceful Uses Atomic Energy, Geneva* **28,** 3 (1958).
8. C. B. Amphlett, L. A. McDonald, J. S. Burgess and J. S. Maynard, *J. Inorg. Nucl. Chem.* **10,** 69 (1959).
9. C. B. Amphlett, L. A. McDonald, P. Eaton and A. J. Miller, *J. Inorg. Nucl. Chem.* **26,** 297 (1964).
10. J. Harkin, G. H. Naneollas and R. Paterson, *J. Inorg. Nucl. Chem.* **26,** 305 (1964).
11. J. I. Bregman, *Ann. N.Y. Acad. Sci.* **57,** 125 (1954).
12. W. J. Maeck, M. E. Kussy and J. E. Rein, *Anal. Chem.* **35,** 2086 (1963).
13. H. J. Schroeder, *Radiochim. Acta* **1,** 27 (1962).
14. J. van R. Smit, U.K.A.E.A. Report AERE-R 3700 (1963).
15. J. van R. Smit, *Nature* **181,** 1530 (1958).
16. V. Kourim, J. Rais and B. Million, *J. Inorg. Nucl. Chem.* **26,** 1111 (1964).
17. A. E. Marcinkowsky, K. A. Kraus, H. O. Phillips, J. S. Johnson Jr. and A. J. Shor, *J. Amer. Chem. Soc.* **88,** 5744 (1966).
18. G. H. Nancollas and R. Paterson, *J. Inorg. Nucl. Chem.* **22,** 259 (1961).
19. C. B. Amphlett and L. A. McDonald, previously unpublished results presented in *Inorganic Ion Exchangers*, Elsevier, London, 1964.
20. C. R. Gardner and R. Paterson, unpublished results.
21. D. E. Conway, J. H. S. Green and D. Reichenberg, *Trans. Faraday Soc.* **50,** 511 (1954).

FURTHER READING

C. B. Amphlett, *Inorganic Ion Exchangers*, Elsevier, London, 1964. (An excellent and comprehensive text.)

Appendix

THE THERMODYNAMIC TREAMENT OF ION EXCHANGE EQUILIBRIA

Chemical equilibrium is best described in terms of *chemical potential*, sometimes known as *partial molar free energy*.

The conditions for and equilibrium distribution of a chemical species, A, between two phases is that the chemical potentials of A in each phase are equal

$$\mu_A \text{ (phase 1)} = \mu_A \text{ (phase 2)} \qquad \text{(A.1)}$$

If in initial stages μ_A (phase 1) is greater than μ_A (phase 2), the system will spontaneously move towards equilibrium by a process of diffusion until by transfer of A from one phase to the other the chemical potentials become equal. The 'driving force' for the diffusion process is proportional to the difference in chemical potentials of the species diffusing. This aspect is more fully developed in Chapter 5.

The chemical potential contains all terms which contribute to the total energy and may be expressed as Eqn. (A.2) for isothermal conditions.

$$\mu_A = \mu_A{}^0 + RT \ln a_A + (P - P^0)\tilde{V}_A \qquad \text{(A.2)}$$

$\mu_A{}^0$ is the chemical potential in a standard state under a pressure P^0. It is a constant with much the same validity as the arbitrary choice of sea level for mechanical potential energy. A typical standard state used when discussing water equilibria is pure water solvent at 25°C and under a pressure of one atmosphere. The chemical potential of water in a solution or in an exchanger may then be related to this standard state by way of Eqn. (A.2). The second term, $RT \ln a_A$, contains the energy contributions of concentration and activity coefficient, γ_A, since the activity a_A, is equal to their product

$$a_A = C_A \, . \, \gamma_A \qquad \text{(A.3)}$$

The activity coefficient measures the effect of interactions of A with surrounding solution species upon the free energy of A. R is the gas constant in suitable units, e.g. cal deg^{-1} mole^{-1}, and T the absolute temperature in degrees Kelvin; 0°C = 273·16°K. The final term is the mechanical or pressure–volume energy contribution. P is the pressure existing in the phase and $\tilde{V}_A$ the partial molar volume of A (essentially the volume occupied by one mole of A in the phase). The partial molar volume is not greatly sensitive to applied pressure and is considered a constant in this treatment.

A-1. SOLVENT EQUILIBRIA AND SWELLING PRESSURE

The equilibrium condition must be that the chemical potentials of water in solution and in the exchanger be equal.

$$\mu_{\mathrm{w}_{\mathrm{solution}}} = \mu_{\mathrm{w}_{\mathrm{exchanger}}} \tag{A.4}$$

Taking pure water at pressure P° as standard state, Eqn. (A.4) becomes

$$\mu^0{}_{\mathrm{w}} + RT \ln a_{\mathrm{w}} + (P - P^0)\tilde{V}_{\mathrm{w}} = \mu_{\mathrm{w}}{}^0 + RT \ln \bar{a}_{\mathrm{w}} + (\bar{P} - P^0)\tilde{V}_{\mathrm{w}}$$

therefore

$$RT \ln a_{\mathrm{w}}/\bar{a}_{\mathrm{w}} = (\bar{P} - P)\tilde{V}_{\mathrm{w}} = \pi\tilde{V}_{\mathrm{w}} \tag{A.5}$$

The osmotic pressure, π, is defined as the difference in pressure between the exchanger and the bathing solution. Since π and $\tilde{V}_{\mathrm{w}}$ are always positive, $a_{\mathrm{w}} > \bar{a}_{\mathrm{w}}$. Equilibrium may only be achieved by having a large pressure–volume contribution to the chemical potential in the exchanger phase. Physically, increasing swelling pressure is the only means of coming to equilibrium in the exchanger phase where dilution is limited or impossible.

A-2. IONIC EQUILIBRIUM AND THE DONNAN POTENTIAL

When an ion, A, having valency z_{A} is in equilibrium across a membrane/solution interface the free energy of the ion is the same in each phase.

$$\mathrm{A}^{z_{\mathrm{A}}}{}_{\mathrm{solution}} \rightleftharpoons \bar{\mathrm{A}}^{z_{\mathrm{A}}}{}_{\mathrm{exchanger}}$$

In this case however the ion will have a total free energy which is the sum of its chemical potential plus the electrical energy it will have by being in a medium with an electrical potential, φ. The value of the electrical contribution is $zF\varphi$, where z is the valency; F, the Faraday of electricity and φ the electrical potential in volts. The units of this term are therefore joules/gram ion, and identical with those of chemical potential. The total free energy of an ion may therefore be represented as an electrochemical potential, $\tilde{\mu}$, where

$$\tilde{\mu} = \mu + zF\varphi \tag{A.6}$$

The equilibrium condition in place of Eqn. (A.1) is now

$$\tilde{\mu}_{\mathrm{A}_{\mathrm{solution}}} = \tilde{\mu}_{\mathrm{A}_{\mathrm{exchanger}}} \tag{A.7}$$

From Eqns. (A.2) and (A.6), Eqn. (A.7) becomes

$$\mu_{\mathrm{A}}{}^0 + RT \ln a_{\mathrm{A}} + z_{\mathrm{A}}F\varphi + P\tilde{V}_{\mathrm{A}} = \mu_{\mathrm{A}}{}^0 + RT \ln \bar{a}_{\mathrm{A}} + z_{\mathrm{A}}F\bar{\varphi} + \bar{P}\tilde{V}_{\mathrm{A}} \tag{A.8}$$

Here we have chosen the same standard state for each phase. For other common choices of standard state it is instructive to consult the rigorous papers of Holm.[1]

Re-arranging Eqn. (A.8), an expression for the Donnan potential is obtained.

$$E_{\mathrm{Don}} = (\bar{\varphi} - \varphi) = \frac{1}{z_{\mathrm{A}}F} \cdot (RT \ln a_{\mathrm{A}}/\bar{a}_{\mathrm{A}} - \pi\tilde{V}_{\mathrm{A}}) \tag{A.9}$$

For a cation exchanger, z_{A} is positive, $a_{\mathrm{A}} < \bar{a}_{\mathrm{A}}$ (assuming the dominance of concentration in activity terms) and $\pi\tilde{V}_{\mathrm{A}}$ is positive. The Donnan potential is therefore negative in the exchanger and positive in solution as predicted by the qualitative arguments of section 3.2. If the solution is diluted the swelling pressure increases [Eqn. (A.5)] and a_A is reduced, both effects increasing the Donnan potential, as can be seen from substitution in Eqn. (A.9).

A-3. CO-ION EXCLUSION

Eqn. (A.9) may be applied. For a cation exchanger, z_A (co-ion) is negative while the pressure and the Donnan potential are unaltered for a given bathing solution. The activity term must therefore be positive and so the activity (and hence concentration) of co-ion in solution is greater than in the exchanger.

A-4. ION EXCHANGE EQUILIBRIA

Consider the equilibrium between counter ions A and B on an ion exchanger:

$$a\bar{A}^{z_A} + bB^{z_B} \rightleftharpoons aA^{z_A} + b\bar{B}^{z_B}$$

a and b are the number of gram ions of A and B respectively involved in the equilibrium. The constraint of electroneutrality requires that

$$a \cdot z_A = b \cdot z_B$$

The condition for equilibrium may be applied separately to each ion using Eqns. (A.1) and (A.2).

$$E_{Don} = \frac{1}{z_A F}(RT \ln a_A/\bar{a}_A - \pi V_A) = \frac{1}{z_B F}(RT \ln a_B/\bar{a}_B - \pi V_B)$$

so that

$$RT \ln \left(\frac{a_A}{\bar{a}_A}\right)^{z_A} \times \left(\frac{\bar{a}_B}{a_B}\right)^{z_B} = \pi(z_B V_A - z_A V_B)$$

From Eqn. (A.3) an expression for the selectivity coefficient may be obtained by substitution.

$$\ln K_A{}^B = \ln \frac{(\gamma_B)^{z_A}}{(\gamma_A)^{z_B}} + \ln \frac{(\bar{\gamma}_A)^{z_B}}{(\bar{\gamma}_B)^{z_A}} + \frac{\pi}{RT}(z_B \tilde{V}_A - z_A \tilde{V}_B) \tag{A.10}$$

For a quite general analysis of thermodynamics of ion exchange, including a variety of reference states, the interested reader is directed to the work of Holm.[1]

REFERENCES

[1] L. W. Holm, *Arkiv Kemi* **10**, 151 and 445 (1956).

Index